N. I. MARKOVSKII

# PALEOGEOGRAPHIC PRINCIPLES OF OIL AND GAS PROSPECTING

*Translated from Russian by R. Teteruk-Schneider*
*Translation edited by R. Amoils*

A HALSTED PRESS BOOK

JOHN WILEY & SONS, New York · Toronto
ISRAEL PROGRAM FOR SCIENTIFIC TRANSLATIONS, Jerusalem

Sole distributors for the Western Hemisphere
HALSTED PRESS, a division of
JOHN WILEY & SONS, INC., NEW YORK

**Library of Congress Cataloging in Publication Data**

Markovskiĭ, Nikolaĭ Iosifovich.
    Paleogeographic principles of oil and gas prospecting.

"A Halsted Press book."
Translation of *Paleogeograficheskie osnovy poiskov
nefti i gaza.*
Includes bibliographical references.
    1. Paleogeography. 2. Petroleum—Geology. 3. Gas,
Natural—Geology. I. Title.
QE501.4.P3M3713          553'.28          75-12798
ISBN 0-470-57215-9

Distributors for the U.K., Europe, Africa and the Middle East
JOHN WILEY & SONS, LTD., CHICHESTER

Distributors for Japan, Southeast Asia and India
TOPPAN COMPANY LTD., TOKYO AND SINGAPORE

Distributed in the rest of the world by
KETER PUBLISHING HOUSE JERUSALEM LTD.
ISBN 0 7065 1517 X
IPST cat. no. 22139 3

This book is a translation from Russian of
PALEOGEOGRAFICHESKIE OSNOVY
POISKOV NEFTI I GAZA
Izdatel'stvo "Nedra"
Moscow, 1973

Printed and bound by Keterpress Enterprises, Jerusalem
Printed in Israel

# CONTENTS

# INTRODUCTION

The unprecedented rate and scope of the development of science and technology have made heavy demands on petroleum and natural gas, both as sources of fuel and energy and as the most essential raw materials for the chemical industry. The 1975 economic plan of the Soviet Union calls for an increase in oil and gas production to 480—500 million tons and 300—320 billion cubic meters, respectively.

This far-reaching program presents theoretical and practical geology with the great challenge of keeping the proven reserves ahead of the increasing demands of industry. The problem is especially pressing in long-worked oil-producing regions where the resources are either exhausted or extremely limited.

How and where new deposits will be discovered, and how large they will be, are very important questions. The larger they are and the closer to the processing and consuming centers, the more profitable they will be and the faster they will recoup the expenditures on their exploration. Therefore, the prediction of and methods of prospecting for large pools take on particular significance. Another aspect should also be emphasized. The results of exploration depend to a greater extent on the methods used, the trends adopted and the scientifically established principles applied, than on the instrumental means of investigation. While the introduction of new equipment in geological prospecting entails considerable expenditures, the use of new up-to-date methods does not call for great outlays, and may yield significant economic and geological results, even when existing equipment is employed.

Scientific prediction, based on the knowledge of the distribution patterns of oil and gas in particular environments, is an essential prerequisite for selecting the most suitable exploration trend. The geological information and experience accumulated to date show that the distribution of oil and gas pools and fields obeys a certain regularity which is controlled by both geotectonic and physiographic factors.

Paleogeographic analysis of the accumulation of producing formations points to the regular occurrence of oil and gas pools mainly along ancient seashores. Under certain favorable conditions, large concentrations of hydrocarbons generally form in the deltaic and prodelta deposits. Paleogeographic studies are also very important in that they constitute the basis for the subsequent expansion of exploration work by revealing lithologic and stratigraphic traps which result from facies variation, pinch-outs, or stratigraphic unconformities. Such traps can only be discovered by resorting to paleogeographic reconstruction.

As far back as the beginning of the 20th century, I. M. Gubkin first discovered a new genetic type of oil deposits — lithologic or the so-called "channel" deposits — whose accumulation is governed by lithologic rather than structural factors. He was of the opinion that the accumulation of oil and gas in anticlinal traps is but a particular case of the gravitational theory of the formation of oil deposits. Gubkin's view that tectonics is responsible for the paths of migration of hydrocarbons and the forms of their accumulations, while lithology is the governing factor in the formation of reservoirs which collect and store these accumulations, is gaining worldwide support.

Fields occurring in lithologic and stratigraphic traps are essential for increasing the prospects of new and, especially, long-worked oil and gas areas. It is sufficient to point out that up to 45%, sometimes up to 90%, of the oil in the U.S.A. is produced from such fields.

The present work discusses the importance of paleogeographic reconstruction as part of the overall regional and other investigations on which exploration drilling for oil and gas is based. It will become evident from the text that only by a comprehensive study of the paleogeographic environment in which a productive deposit was formed, paying due attention to other factors affecting oil and gas accumulation, can one expect to obtain better results from prospecting operations.

Special emphasis is placed on the association of the largest oil and gas fields with paleoriver deltas. We have every reason for stressing this point. Such an association has been noted in numerous oil and gas basins throughout the world. It is very significant that a special symposium on "Deltaic Sedimentation and Petroleum Occurrences," held in Canada in 1970 under the auspices of the American Association of Petroleum Geologists, voiced the opinion that lithologic studies, which furnish information for the petroleum geologist, should mainly focus on deltaic sedimentation. The mentioned symposium summed up the vast body of factual material assembled in the U.S.A. and Canada in recent years, indicating the close relationship between large hydrocarbon accumulations and sedimentation in ancient deltas. The experience gained in the Soviet Union supports this conclusion.

In this work we proceed from the principal assumption that lithogenesis and tectogenesis play the key role in the processes starting from the initial deposition and burial of the organic substance and leading up to the accumulation of hydrocarbons in reservoir rocks. According to Strakhov (1971), lithogenesis, which is customarily associated with tectogenesis, cannot be considered apart from the physiographic environment of sedimentation. Only the theory of lithogenesis, given a geographic substantiation, can establish a correlation between various types of lithogenesis and life in the outer geospheres. Strakhov holds that a general idea of ancient lithogenesis at each particular stage of lithological development can be gained from knowledge of Recent sedimentation. Paleogeography is a tool for the comprehensive investigation of the conditions and mechanism of sedimentation, including those relating to oil- and gas-bearing strata.

In recent years petroleum and gas geology has been enriched by new ideas on oil-generating source rocks, the main phase of oil formation (Vassoevich, 1969), the vertical zonality of hydrocarbon accumulations (Vysotskii and Olenin, 1964) and other concepts, which make possible the identification of source rocks in the stratigraphic section. Complementary paleogeographic reconstructions make it possible to establish the zonality of oil and gas occurrences

in plan and to predict which areas will contain the largest hydrocarbon accumulations. Moreover, by establishing the genetic nature of the producing
formations, paleogeographic reconstructions assist one to determine the
optimal location of both producing and test wells.

New concepts and trends will play a significant role in solving the complex problems involved in bringing about a considerable increase in proved
oil and gas resources. The present work is aimed at the popularization and
introduction of one such trend. Since no trend in science is ever absolute
or final but is based on a fairly complete system of knowledge and experience,
the paleogeographic trend in petroleum and gas geology will undoubtedly be
refined in the process of its utilization. But first of all one must be aware
of its essence, its theoretical and practical implications.

Scientific generalization usually leads to scientific prediction which, in
turn, is directed at the solution of practical problems. The author, relying
both on personal observations over many years and published sources, has
undertaken to demonstrate the potential of paleogeographic studies in oil and
gas exploration.

# Part One
# GENERAL PRINCIPLES

## 1. STRUCTURAL-TECTONIC AND PALEOGEOGRAPHIC APPROACH TO OIL AND GAS EXPLORATION

None of the factors involved in the formation of commercial oil and gas deposits is sufficient in itself, as this process is promoted only by a certain combination and interplay of these factors. Nevertheless, some of them have a key function in determining the character of the process. One such factor, in addition to geotectonics, is the physiographic environment in which the producing formations were formed. This implies a combination of conditions, such as the geographic setting of the locality at a certain time, the climate, the extent of development of organic life, the facies and geochemical features of sedimentation, etc. Therefore, paleogeographic conditions, encompassing several of the most important premises of normal sedimentogenesis, may be listed among the principal factors exerting a decisive influence on the formation of oil and gas deposits.

Prediction and the solution of specific problems encountered in exploration work are not feasible if we lack information on the former location and character of the ancient seas and continents, the coastal zones, the rivers and their deltas. If one is unable to picture the paleoclimate, the direction of transport of terrigenous material, the interrelationships between the producing and enclosing formations and the optimal environments for oil and gas accumulation, and if one is unaware of how these physiographic conditions varied in time and space, one should not anticipate promising results. This is why paleogeographic reconstructions are no less important than structural-tectonic ones. The two trends should be applied in combination in research relating to oil and gas geology. This will make it possible to gain a deeper insight into the distribution patterns of oil and gas deposits and to make the most of them.

The efficacy and economic feasibility of exploration work are largely dependent on the level of scientific substantiation, which, in turn, is determined by the particular concept or hypothesis underlying the prediction of the presence of oil and gas in a particular area. Scientific prediction is the foundation of successful prospecting for oil and gas deposits.

It cannot be denied that Soviet geologists have prepared a number of forecast maps covering various zones, regions, and even the entire territory of the U.S.S.R. However, in most cases these maps are too general and cannot serve as a basis for selecting a particular exploration method.

Until recently, one of the main criteria in prospecting was the presence of favorable local structures in a potential or proved oil and gas basin. Several new indicators for exploration work have become available in the present decade. They include advances in the theory of lithogenesis, further development of theoretical assumptions pertaining to petroleum geology, geochemistry and hydrogeology, historical, geological and geochemical methods of assessing the oil and gas content in sedimentary basins, etc.   Today it is possible to detect strata with a higher than average content of organic material, and there are even means of determining the degree of catagenetic transformation of this material.

All these achievements undoubtedly contribute to a more objective evaluation of oil and gas potentials and serve as a guide in selecting the most efficient exploration methods.   Nevertheless, at least 60 to 70% of all test wells are dry holes, and this is a most convincing reason to seek better criteria making possible prediction on a more sound scientific basis, i. e., taking into account all the characteristics of and regularities governing the formation and occurrence of oil and gas deposits.

Drilling through nonproducing structures is unavoidable until more direct methods of detecting oil and gas pools are devised.   However, with a more rigorous scientific approach to the choice of local structures for deep exploration, the element of chance will be reduced to a minimum.   Even in the West Siberian basin, which is extremely rich in oil and gas, the efficacy of exploration work can be further improved with the aid of more accurate predictions based on structural-tectonic criteria, as well as other, no less important considerations.

The experience gained over more than a century of oil and gas prospecting has shown that the theory of the organic origin of oil and gas is best suited to practical application.   Practice has shown that the distribution of hydrocarbon deposits is closely related to the features of their biogenesis.   This makes it possible to assess the prospects of a certain territory and the distribution patterns of commercial oil and gas accumulations, and to organize the search for both large fields as well as lithologic and stratigraphic pools.

We now have at our disposal an enormous body of information on the tectonics of the main oil and gas fields, and on the composition, thickness, facies, physical properties and other characteristics of the sedimentary rocks. Synthesis of these data can and should be used to establish the location and distribution of the source rocks in which the oil and gas were originally generated and of the reservoir rocks, and to determine the most promising trends and rational techniques to be employed in exploration work.   To this end, it is necessary, primarily, to reconstruct the sequence of physiographic environments responsible for the formation of individual producing formations or their analogues.   Even the most detailed structural map, if lacking a paleogeographic foundation, is not able to serve as an objective basis for selecting structures for deep exploration.

Gubkin (1940), in his well-known work on the Ural-Volga oil field, pointed out that American geologists, after they have detected most of the structures, start searching for stratigraphic traps associated with unconformities.   He emphasized that this calls for a most meticulous study of the paleogeography: the changes in the location of seas and continents with time, i. e., a comprehensive investigation of the entire geological history of a given section of the earth's crust.

This method has helped to reveal the largest oil regions, e. g., those in East Texas where oil accumulated in the upper members of the Upper Cretaceous sediments unconformably overlain by younger rocks.   Gubkin was convinced that Soviet geologists should also conduct paleogeographic studies.

More than 30 years have passed since these ideas were first expressed. It is now evident that structural-tectonic studies of oil and gas deposits must be supplemented by a thorough analysis of the physiographic environment in which each producing horizon was formed.

A characteristic feature of current exploration work is its virtual restriction to the detection and drilling of anticlinal deposits; moreover, there are no strict criteria for selecting the structures to be investigated. Hundreds of local uplifts are known to exist in certain oil and gas basins, while in others all the structures have been explored. In the former case it is uncertain which of these uplifts merits preferential treatment in exploration work, and in the latter case it is not clear where new pools, unassociated with structural traps, should be sought. Geological science is now confronted with these practical problems.

It is now high time to bring both prediction and oil exploration into conformity with current scientific advances.

New criteria for the assessment of both old and new oil- and gas-bearing territories will be forthcoming from paleogeographic studies and these should be utilized to the full.

*Paleogeography: Importance and Place in Oil and Gas Geology*

While giving due credit to the other factors influencing the formation of oil and gas pools, we shall attempt to define the place and role of paleogeography among them.

Differentiation and specialization are known to be characteristic trends in modern science. The tendency toward subdividing oil and gas geology into constituent branches facilitates the successful solution of practical problems. In fact, whereas geotectonics deals with structural features of the earth and their variation and evolution as influenced by deep-seated processes, paleogeography is the discipline dealing with the movement of matter on the earth's surface under the effect of physiographic and biogeographic, including also paleoclimatic, factors. In other words, paleogeography is the science which studies the configuration and characteristic features of the ancient seas and continents, which ultimately determined the character of sedimentation in the past. It is closely related to the theory of facies, lithology, geochemistry and other branches of geology, as well as to geotectonics. According to Rukhin (1959), paleogeography is a geological science which studies the physiographic conditions which existed on the earth's surface in past ages.

We need not dwell on the concept of paleogeography as a discipline or on its position among related fields, since these aspects have been given sufficient coverage in the well-known works of A. P. Karpinskii, A. A. Borisyak, Yu. A. Zhemchuzhnikov, D. V. Nalivkin, N. M. Strakhov, B. L. Lichkov, L. B. Rukhin, V. I. Popov and others. Numerous paleogeographic problems have been successfully tackled by V. P. Baturin, B. P. Zhizhchenko, G. F. Krasheninnikov, S. G. Sarkisyan, G. A. Smirnov, V. A. Grosgeim, T. I. Gurova, V. P. Kazarinov, K. S. Maslov, O. L. Einor and others.

However, notwithstanding the obvious achievements in this field, the available paleogeographic maps are usually schematic, merely showing the distribution of various lithologic formations or mineralogic-petrographic associations. Even in those cases where they do depict the regions of ancient seas and land areas, they very seldom specify the sea depths, the type of salinity of undercurrents, or, in the case of land areas, the features of the topography and the pattern of lakes and swamps, river valleys and deltas. Existing paleogeographic maps scarcely differentiate between paleoclimatic and biogeographic zones. All these missing details are essential for the prediction of and exploration for commercial accumulations of hydrocarbons.

Paleogeographic studies may be extremely helpful in reconstructing the geologic evolution of a given territory, in revealing the genetic features of the distribution of the sedimentary deposits in which oil and gas were originally formed, and in establishing the distribution patterns of hydrocarbon deposits.

In order to gain an understanding of the conditions of formation and the distribution patterns of oil deposits, it is essential to establish a connection between the latter and the productivity of different biosphere media. This cannot be done without resort to paleogeographic reconstruction.

Paleogeographic reconstruction alone obviously cannot solve all the complex problems encountered in oil and gas exploration. The strong point of science, however, is the interlinking of its various branches. It is therefore necessary to assess the proper place and role of paleogeographic studies vis-à-vis tectonics, lithology, stratigraphy, geochemistry and other important disciplines. This can only be done by generalizing practical experience and its theoretical implications.

The majority of oil and gas fields occur in arenaceous and sand-silt rocks which possess granular porosity and intergranular permeability. According to Khanin (1969), oil and gas pools confined to terrigenous reservoir rocks account for 74% of the total reserves in the Soviet Union, while those associated with carbonate reservoirs constitute 18%, and those in terrigenous-carbonate reservoirs make up 8%. More than 80% of the gas and gas-condensate pools occur in sand-silt sediments. Nearly all oil and gas fields of the rich West Siberian basin are concentrated in sand-silt rocks.

In 236 major oil fields of the capitalist countries 59% of the reserves are located in sands and sandstones, 40% in limestones and dolomites, and 1% in fractured shales and weathered metamorphic and igneous rocks. If 21 Near and Middle East deposits with oil occurring in carbonate rocks are excluded from the mentioned 236 fields, the reserves of the major deposits of the capitalist countries are distributed as follows: about 77% in sands and sandstones, 21% in limestones and dolomites, 2% in other rocks.

The above statistical data set one thinking about the causes of such a distribution. It is not sufficient to claim that terrigenous rocks have a higher porosity and permeability than carbonate rocks. The point is that not all terrigenous rocks, even those with high porosity and permeability, contain oil and gas. Oil and gas accumulations are known to be formed only in reservoirs sealed by impermeable rocks forming tectonic or lithologic traps. However, even this does not complete the picture, since test wells fail to encounter oil in many structures which can be considered as perfect reservoirs in terms of their shape, size and sealing.

Consequently, there are other factors, aside from those mentioned above, which control the distribution of oil and gas pools; they are either yet unknown or have been disregarded. Soviet and, in particular, non-Soviet experience suggests that one such factor is the paleogeographic conditions under which zones and individual areas of oil and gas accumulation were formed. In point of fact, the presence of organic matter is a prerequisite for the onset of the complex and prolonged process leading to the formation of hydrocarbons and their deposits. Uspenskii (1970) lists the following principal factors governing the concentration of organic matter in sediments:

1) the amount of living organisms (biomass) in the sphere of accumulation of the given sediments — on the ground surface and in the overlying water layer;

2) the intensity of decay of organic matter, which may be expressed by the fossilization coefficient (the ratio of buried organic matter to total organic matter);

3) the relationship between the influx of organic matter into the sediments and the introduction of mineral components;

4) the extent and rate of mechanical transport of the organic detritus from certain facies environments to others;

5) the local displacement of the mobile components of organic matter within a sediment undergoing compaction in the early diagenetic stage.

Consequently, the distribution of the organic matter, which is the starting material for the formation of oil and gas, in the sediments is directly related to the physiographic conditions controlling their deposition. It has been established that higher concentrations of organic matter are characteristic of argillaceous rocks. Sapropel varieties of organic matter are common in subaqueous sediments which are deposited under conditions of maximum isolation from the influence of land areas, while humic varieties are appreciably affected by dry-land conditions.

Hydrocarbon deposits are mainly associated with the zone transitional from dry land to sea, where the conditions are favorable for the development of organic life and where terrigenous deposits accumulate. The influence of the physiographic environment of sedimentation on the distribution of both organic matter and reservoir rocks is discussed below.

*Non-Soviet Experience*

One may draw on the experience of the U.S. Geological Survey in order to assess the role of paleogeographic substantiation in exploration work. For instance, drilling data for the Appalachian oil and gas basin are presently being reassessed on a paleogeographic basis with a view to detecting additional lithologic and stratigraphic deposits. The paleogeographic reconstructions in this case constitute the basis for directing exploration activity. Paleogeographic studies are being conducted on a particularly wide scale in the Illinois basin. In the Mid-Continent region the directions of the river valleys in Pennsylvanian times are being determined by the paleogeographic method, while in Kansas this method is used for tracing the buried topography of pre-Mississippian deposits.

The location of wells in the Delaware basin is determined with the aid of paleogeographic reconstructions. All the available data on the Gulf Coast Province are being given a paleogeographic reassessment to prepare a new drilling program. Pinch-out zones are being sought in the Rocky Mountains Province. Production from the long-worked Williston basin is now being recommenced thanks to paleogeographic studies which have revealed a large number of new oil and gas deposits of the lithologic type. In the Bighorn and Wind River basins paleogeography is the principal tool in exploration for pools of this type. Detailed seismic studies supplemented by paleogeographic analysis have detected several lithologic deposits associated with sand lenses in the Powder River, San Juan and Uinta basins and in other regions.

The producing Cherokee sandstones in the central part of Oklahoma are channel-shaped, being the result of the filling of pre-Pennsylvanian erosional valleys with fluviatile deposits.

In 1969 Silurian oil-bearing sandstones, representing deposits of small deltas along the coast of an ancient shallow sea, were discovered at depths ranging from 1200 to 1600 m in northeast Ohio. A large oil pool formed in deltaic sand deposits was discovered in the East Canton Magnolia field.

The fluviatile sandstones of the Muddy River are the main object of oil exploration in the Rocky Mountains. The reserves in the Bell Creek field alone amount to more than 27 million tons. In addition to fluviatile deposits, the sandstones of ancient barrier beaches are also productive. Many more examples could readily be furnished, but those cited serve as sufficient proof of the decisive role played by paleogeographic studies in exploration for oil deposits occurring in lithologic and stratigraphic traps, which account for about 45% of the overall oil output in the U.S.A. American geologists believe that this figure will reach 75% in the foreseeable future. Such well-known oil fields as those in East Texas, Pembina, Seeligson, West Edmond, Oklahoma and a number of others are mainly controlled by lithologic, rather than structural factors.

Ninety-five percent of the world's known oil and gas pools associated with lithologic and stratigraphic traps are in the U.S.A. Such pools in the Soviet Union account for no more than 6 to 7% of the annual oil production; they were discovered by drilling in local structures.

If one traces the history of oil exploration in the U.S.A., one readily notes that the search for lithologic and stratigraphic pools started once all the local structures had been explored. Thus, new pools are being discovered in the following basins which were thought to be thoroughly investigated and well covered by drilling: the Appalachian, Illinois, Delaware, Williston, Powder River, San Juan, Uinta, etc. The new oil fields and pools discovered in the above basins are generally controlled by the lithologic factor, the characteristic features of which are being revealed with the aid of detailed paleogeographic studies.

We should learn from the American experience, rather than wait until all local uplifts located in a particular oil and gas basin are finally explored. Reconstruction of physiographic environments and exploration for oil and gas pools associated with all types of traps should parallel exploration of local structures.

Although the search for stratigraphic and lithologic traps is common
practice in the U.S.A., there is no special exploration technique and the location of boreholes is usually according to the wildcat principle.  However,
this is not our prime concern.  We are mainly interested in the general
distribution patterns of oil and gas pools, characteristic of all traps.  Since
the territory of the U.S.A. is essentially covered by boreholes, and its oil-
and gas-bearing regions have been adequately studied, it might be useful to
take a closer look at some of them.

The base of the sedimentary cover of the North American plate mainly
consists of Cambrian and Ordovician rocks.  Seven regional breaks were
detected in the cover proper.  The most significant structural changes took
place during the Pennsylvanian and Permian, as well as in the Mesozoic.
The basement shows a general subsidence to the south of the Canadian Shield.
The intensity of tectonic movements increases in the same direction.  The
East European Platform is also characterized by an increase in general
subsidence and intensity of tectonic processes in the direction from the Baltic Shield to the Ural geosyncline.  Systems of deep faults and the presence
of Permian salt-bearing syneclises (Caspian in the U.S.S.R. and Permian in
the U.S.A.) are typical of both regions.

The Mexican syneclise, measuring about 2 million $km^2$ in area and situated in the southeastern part of the country, is one of the richest oil- and
gas-bearing regions in North America.  It is a region of persistent and
steady subsidence and sedimentation.  Within it is located the Gulf Coast
trench, up to 200—250 km wide and extending latitudinally for 1250 km;  the
basement is at a depth of 15 to 18 km.

The huge oil- and gas-bearing Gulf Coast basin encompasses the coastal
plain in the south and southeast of the U.S.A. and also the continental shelf
of the Gulf of Mexico.  Hydrocarbon deposits occur in pinch-out zones in
arenaceous series.  One third of all U.S. oil is produced here.  In all stratigraphic subdivisions, from the Mesozoic up to the present, there is a gradual
transition from rudaceous sediments to arenaceous-argillaceous, argilla-
ceous-calcareous and carbonate deposits along the dip.  A major role in the
Paleogene section is played by arenaceous-argillaceous fluviatile, deltaic,
prodeltaic and nearshore-marine formations which give way downdip to
deep-water marine facies.  The Neogene section consists of a characteristic
complex of fluviatile and deltaic deposits whose thickness exceeds 6000 m
in the coastal zone of the Gulf of Mexico.  Further south they are replaced
by thick series of clay and marl with intercalations of marine sand.

The rather deep and vast depression with the Gulf of Mexico at its center
is one of the richest oil and gas basins in the world.  There are no foredeeps
or piedmont troughs along its margins, which makes it somewhat similar to
the West Siberian oil and gas megabasin, although the maximum downwarping
of the latter is only one third that of the former.

The general subsidence of the Mesozoic and Cenozoic rocks toward the
Gulf of Mexico is complicated by several regional faults and flexures.  Salt
tectonics is also quite common in this region.

Commercial accumulations of oil and gas are encountered throughout the
stratigraphic column — from the Jurassic to the Quaternary.  Insignificant
amounts of oil and gas are produced from Paleozoic horizons.  The Cenozoic

deposits account for more than 50% of the total output.  The fields are generally multilayered, sometimes containing more than 30 producing sand beds with argillaceous intercalations.  The oil and gas are encountered at depths ranging from several dozens of meters to 5 or 6 km.  However, the main pools and reserves are located at depths ranging from 600 to 1800 m in the northern regions and from 1200 to 3000 m in the southern regions.

The oil and gas in the richest producing regions occur in structural-stratigraphic and structural-lithologic traps.  The pools in the major fields are often associated with unconformities.  For instance, the unique field in East Texas is associated with a zone of discordance between the producing Woodbine sandstone and the underlying older strata.  The oil pool there is sealed by shales unconformably overlying the sandstone.  This pool is 112 km long, 8 to 20 km wide and up to 300 m thick.  The producing formation is clearly of alluvial-deltaic origin.

Pools of the lithologic type occurring in sand lenses within argillaceous strata are very common in Jurassic and Early Cretaceous formations to the north of Sabine-Monroe and in south Arkansas.  Principal production in these regions is from sand lenses of the Gulf series (El Dorado, Smackover, Schuyler and other fields).  Lithologic pools in sand lenses are also known in East Texas (Long Lake, South Payne, etc.).  The gas pools in the Bethany field are located in sand lenses at fifteen different stratigraphic levels.

The South Texas region contains large zones of oil and gas accumulation, associated with the regional pinch-out of sand formations along flexures and faults.  These sands were deposited in the environment of a regressing marine basin, so that the pinch-out zones of the producing formations run parallel to the Gulf Coast.  For instance, the Anáhuac and Frío producing formations form a band 65 to 90 km wide and 1200 km long, extending from the Mexican border to Mexico state.  A similar zone of Miocene age is situated further south.  Numerous pools are associated with sand lenses of the fluviatile type and with sandbars.

Oil and gas pools in the Rocky Mountains basin (Williston, Bighorn, Powder River, Wind River, Laramie, Denver, Paradox, San Juan) are distributed mainly on the rims of piedmont troughs and depressions.  Drilling operations in anticlinal structures, which were started in 1912, revealed the presence of minor pools.  Only in the postwar period, with the onset of exploration for lithologic and stratigraphic traps and deposits located on the flanks of buried structures, were large deposits also discovered.  Eight major deposits account for 35% of the oil produced in the region.

All the natural oil-bearing reservoirs in the Powder River basin are sandstones, the abundance of which depends on the primary physiographic conditions of sedimentation.  The producing sandstones were deposited on the shelf and are nonpersistent.  Carbonate rocks are common in the central part of the basin.  Oil occurrences are confined to the shelf slopes, in particular to those areas where the amount of arenaceous sediments sharply increases.

The Denver basin, whose area is 150,000 km$^2$, is composed of rocks ranging in age from Cambrian to Recent, with an overall thickness of up to 4500 m. Oil is mainly produced from the Early Cretaceous Dakota series — interbedded sandstones and shales transgressively overlying Jurassic rocks;

218 oil pools were discovered in the Dakota sandstones.  The arenaceous-argillaceous Dakota formations are represented by shallow marine and coastal terrestrial facies which accumulated as the result of several transgressive cycles of sedimentation.  The transgression was interrupted by periods of slight regression.  The most significant traps in the Denver basin are those resulting from the pinch-out of the sandstones updip.  Very few of the oil fields are located in anticlinal uplifts.  An improved technique of seismic surveying is used for detecting structural details and lithologic changes.

Oil and gas pools in the Appalachian basin are classified in four categories: in extensive sand horizons; in nonpersistent sandstones or lenses; in carbonate reservoirs; in argillaceous beds.  We shall consider the first two categories.

According to Woodword (1958), the extensive sand horizons include the Lower Silurian, Lower Devonian (Oriskany formation) and Lower Mississippian (Lower Carboniferous) sandstones.  Most oil and gas is produced from the Devonian and Silurian sandstones in the marginal zone of the basin, where they pinch out.

The hydrocarbons migrated updip from the central portions of the basin. The linear zones of increased porosity extend along the western boundary of the basin; within ancient river beds and shoals they are perpendicular to the direction of the regional decrease in thickness.  In these reservoirs the structural factor was of minor significance in the formation of primary accumulations.

The above also applies to the underlying sand horizons of the Cambrian, which have hardly been explored by test wells to date.

Considerable amounts of oil and gas are produced from nonpersistent sand horizons, i. e., from those forming irregular or lenticular bodies, in the Appalachian basin.  The sand lenses are relatively small and are separated from one another within argillaceous, in places carbonate, formations. Most reservoirs of this type occur in the ancient shoal region.  Many of them grade into terrestrial facies.  Nonpersistent sand horizons are most common in the Upper Devonian.

The spatial distribution of these horizons and the associated oil and gas conform to a certain regularity: they occur in the zone of transition from coarse-grained terrestrial to argillaceous marine sediments.  The structural, more specifically, geotectonic factor played a minor role during the primary formation of these pools.  Hydrocarbon accumulations in elongated linear sandbar deposits or in reefs are oriented parallel to the ancient shoreline.  The pools in the fluviatile deposits are perpendicular to the ancient shoreline.

Among the giant fields, defined by American geologists as those with total reserves exceeding 80 million $m^3$ of oil or 95 billion $m^3$ of gas, there are quite a few whose formation was largely governed by the lithologic factor.  According to King (1971), these include the following:

OIL   FIELDS

|  | Reserves in millions of m$^3$ |
|---|---|
| Bolivar Coastal (Venezuela) ..................... | 4700 |
| Prudhoe Bay (Alaska, U.S.A.) ................... | 3180 |
| East Texas (Texas, U.S.A.) ..................... | 954 |
| Pembina (Alberta, U.S.A.) ...................... | 286 |
| Quirequire (Venezuela) ......................... | 159 |
| Boscan (Venezuela)   ........................... | 159 |
| Karamai (Chinese Peoples Republic) ............. | 111 |
| Old Illinois (Illinois, U.S.A.) .................. | 95 |
| Mene Grande (Venezuela) ....................... | 95 |
| Kern River (California, U.S.A.)  ................ | 95 |
| Bradford (Pennsylvania, U.S.A.)  ................ | 95 |

OIL   FIELDS   IN   REEFS

| Poza Rica (Mexico) ............................ | 318 |
|---|---|
| Kelly-Snyder-Diamond (Texas, U.S.A.) .......... | 254 |
| Naranjos-Cerro Azul (Mexico)   ................. | 222 |
| Intisar "A" (Libya)  ........................... | 222 |
| Swan Hills (Alberta, Canada) ................... | 206 |
| Intisar "D" (Libya)  ........................... | 190 |
| Rainbow (Alberta, Canada) ..................... | 111 |
| Redwater (Alberta, Canada) .................... | 111 |
| Dahra-Hofra (Libya)  .......................... | 111 |
| Leduc-Woodbend (Alberta, Canada) ............. | 95 |

GAS   FIELDS

|  | Reserves in billions of m$^3$ |
|---|---|
| Hugoton (Kansas-Oklahoma-Texas, U.S.A.) ....... | 1960 |
| Blanco Basin (New Mexico, U.S.A.) ............. | 308 |
| Jalmat   (New Mexico, U.S.A.)   ............... | 224 |
| Mocane-Laverne (Oklahoma, U.S.A.) ............ | 152 |

When discussing the fields occurring in lithologic and stratigraphic traps, King emphasized that they may be present in great numbers in basins which are characterized by unstable tectonic evolution, large stratigraphic uncon-formities and transgressive overlaps, as well as by alternation of paralic marine and fluviatile facies.  This is another reason why paleogeographic studies are important for developing methods of exploring for pools asso-ciated with traps of this type.

The Midcontinent region in the U. S. A. is unique in the variety of its lithologic and stratigraphic traps.  By the beginning of 1968, 2101 oil and gas pools in the state of Oklahoma alone were known to be associated with these particular types of traps, while structural traps contained 1182, and unclassified traps — 93.  The total hydrocarbon reserves in nonstructural traps are believed to exceed those in the structural pools.

The Powder River basin is extremely rich in petroleum deposits of the nonstructural type, which occur in sandbars and shoestring sandstones of Cretaceous age, as well as in Mississippian carbonate deposits truncated by a stratigraphic unconformity.  The traps in the Gulf Coast region are confined to numerous paleodeltas of Eocene age, stretching across the Gulf Coast, Texas, Louisiana, and southwest Mississippi.

There is no need to furnish additional examples here, as we intend to discuss some of them later on.

To conclude this brief review of the distribution of oil and gas deposits in the U. S. A., we wish to point out that the prospects of discovering deposits of any type, especially those associated with nonstructural traps, are far better in the Soviet Union.  When comparing the East European and North American platforms, the Caspian synecline attracts particular attention.  It has an area of some $560,000 \, km^2$, and the maximum depth of downwarping of the basement exceeds 10 km.

This synecline is filled with a thick series of sedimentary formations, mainly terrigenous and hydrochemical.  The best-studied areas are the marginal portions.  For instance, quite a few oil fields in the Volga-Ural oil and gas region are located on the northwestern flank, which also consti- tutes the southeastern slope of the East European Platform.  However, the true resources of this enormous synecline have yet to be explored.  This region may be compared to the rich Mexican basin which is sometimes re- ferred to as the "world pole of oil and gas" in the Western Hemisphere. The Caspian synecline may well prove to be its counterpart in the Eastern Hemisphere.  We are convinced that paleogeographic studies will play an essential role in the comprehensive investigation of this region.

## 2.  STRUCTURE OF THE BIOSPHERE AND PALEOGEOGRAPHY

Suess (1875) introduced into geology the concept of the biosphere as that part of the earth's crust containing some forms of life.  At the beginning of this century Vernadskii, the founder of biogeochemistry, developed the theory of the biosphere, which is especially well suited to the purposes of geochemistry, physiography, and petroleum geology.  He defined the term "biosphere" as the "realm of life," which may manifest itself only in a certain medium and under certain physicochemical conditions.

Vernadskii's theory of the geological role played by living organisms brought about a true revolution in the natural sciences.  He pointed out that living organisms are a powerful geological force.  "There is no chemical force on the earth's surface which is more persistent and therefore more potent in its ultimate consequences than living organisms taken as a whole. . . In fact, living matter exerts an influence on the overall chemistry of the earth's crust and controls the geochemical history of nearly all the elements constituting the crust" (Biosfera, 1967, pp. 241 and 127).

According to Vernadskii, four modes of occurrence of chemical elements can be distinguished in the earth's crust:  in rocks and minerals, in magma, as dispersed elements existing in the free state, and in living matter.  The chemical elements, once they enter a living organism, find themselves in a special medium "which has no analogs anywhere else on the earth." Such an assessment of the role played by living organisms provides a proper approach not only to the problem of oil formation, but also to the processes leading to the formation of oil pools. This is the reason why paleogeographic reconstructions of the remote past have to be considered in conjunction with the specific features of the distribution of life.

With the appearance of life on earth the biosphere came into being, within which complex chemical transformations take place, generated and supported by living organisms which are intimately connected with their environment. According to Vernadskii, "There is not a single organism which is not connected, at least partially, with inert matter by the process of its feeding and respiration."

All organisms are classified either as autotrophs, i. e., not reliant on other organisms for their nutrition, or as heterotrophs, i. e., feeding on organic compounds produced by other living organisms. Unlike the heterotrophs, the autotrophs develop entirely from inorganic matter. Nitrogen, oxygen, carbon and hydrogen, which constitute the bulk of their composition, are derived from mineral matter.

There are two distinct groups among the autotrophs: green chlorophyllous plants and rapidly reproducing bacteria. The chlorophyllous organisms, including the green plankton of seas and oceans, are the major biosphere mechanism which produces, by way of photosynthesis, chemical bodies that accumulate radiant solar energy. Bacteria, which are very common in water bodies and their sediments, possess a geochemical energy which is tens or even hundreds of times greater than that of plants.

The most important feature of these autotrophic bacteria is their global distribution. However, the absence of large concentrations of such organisms is due to the special conditions of their nutrition. They derive energy from the oxidation of compounds of nitrogen, sulfur, iron, manganese and carbon. These compounds are usually more or less oxidized before the bacteria start working on them. Therefore, the amount of nutrients available to the bacteria is rather limited. There exists a balance between sulfate-reducing bacteria and sulfate-oxidizing autotrophic organisms.

From the moment of origin of life and throughout the subsequent geological ages, the distribution of living organisms in the hydrosphere was generally maintained in particular zones, notwithstanding the variability of life forms and of the location of the hydrosphere. Life in seas and oceans is mainly concentrated in regions where maximum transformation of solar energy takes place. Vernadskii, while considering the most general aspects of the distribution of life in the ocean, distinguished four static agglomerations of life: two films — plankton and bottom, and two concentrations — nearshore and sargasso.

The nearshore concentrations sometimes encompass the entire water layer, including the bottom film. They are always associated with the shallower parts of the ocean, seas and coastal regions. The bottom film is the region where the chemical activity of life is manifested strongly; it is a reducing medium and is the site of continuous deposition of dead organisms derived from the plankton film and nearshore concentration. These organic remains teem with anaerobic bacteria which contribute to the reducing nature of the medium.   The bottom films together with the adjacent nearshore concentrations of life seem to present the most favorable environment for the accumulation of prospective petroleum-generating deposits.

The distribution of life is different on dry land where the soil and its flora and fauna in fact constitute a single living film.   However, even within this film there are concentrations of living matter in various inland water bodies.   Soil is a predominantly oxidizing medium and provided it is moist, life develops in it.   The presence of water is an essential prerequisite for the existence and reproduction of living organisms.   The bulk of the water in the terrestrial biosphere is concentrated in swamps and lakes where peat and sapropel accumulate.

Consequently, the above-mentioned films and zones of concentration of living matter are regions where most diverse systems of dynamic equilibria of the chemical elements occur, and where radiant solar energy assumes the form of free chemical energy.   The existence of these regions depends on the penetration of the sun's rays, as well as on the properties of the living matter which accumulates the solar luminous energy and transforms it into chemical energy.

It is quite obvious that there is a direct connection between the distribution of seas and oceans, on the one hand, and the nearshore concentrations of life, on the other.   This relationship undoubtedly plays a decisive role in the process of oil and gas accumulation, and may be elucidated only by resort to reconstruction of the paleogeography of particular periods of evolution of the earth's crust.

The theory of the biosphere was further developed in the works of A. P. Vinogradov, N. M. Strakhov, B. B. Polynov, V. I. Lebedev, N. B. Belov, A. I. Perel'man and others.   Vernadskii's ideas on the geochemistry of oil and gas are being successfully developed by N. B. Vassoevich, V. V. Veber, A. A. Kartsev, S. G. Neruchev, O. A. Radchenko, K. F. Radionova, V. A. Sokolov, V. A. Uspenskii and others.   Uspenskii (1970) believes that the bulk of living matter in the biological structure of the biosphere is contained "in the main foci of life where the evolution of life is extremely intensive and where the world of autotrophic and heterotrophic organisms is in equilibrium because the processes of synthesis and decomposition are balanced."   He furnishes data to illustrate the tremendous role played by living matter in the process of formation of the earth's sedimentary cover.   The total amount of living matter produced annually constitutes about 0.003% of the sedimentary mantle. However, the total amount of this material produced throughout the entire period of existence of the biosphere is about 30 times greater than that of the present mass of sedimentary rocks.   Only a negligible part of the organic matter is preserved in the sediments.   A mere 8.7% of the total carbon in ocean waters is in the form of dissolved organic matter.   The remaining 91.3% is accounted for by fixed carbon and dissolved carbon dioxide.

The evolution of life and accumulation of organic matter are most intensive in the nearshore parts of oceans and deep-water seas.   Much organic matter transported from the land areas accumulates on the shelves and in shallow seas, bays and lagoons, alongside local plant and animal organic matter.   The decay of this organic matter gives rise to a reducing environment uncharacteristic of continental and deep-water sediments.   Regions

with persistent reducing environments are therefore confined to intermediate epicontinental zones of the hydrosphere, i. e., to epicontinental seas
fringing continental blocks and to inland water bodies.  The total area occupied by these environments (according to Uspenskii) amounts to 15—20% of
the earth's surface.  They are the sites of maximum sedimentation and
maximum concentration of organic matter, where all mineral fuels originate.

The replacement of marine conditions by continental ones leads to a re-
working of the sediments and to decomposition of the organic matter contained in them.  However, in each new cycle of sedimentogenesis new organic
material is produced by biogenic synthesis.  According to Uspenskii, a rough
calculation gives enormous values for the total mass constituted by the living
matter which has populated the earth since the beginnings of life.  This mass
considerably exceeds that of the present-day oceans.  More than 70% of all
the sedimentary material deposited in the oceans remains in the epicontinental shelf zone, while only 14% finds its way to the deep-sea areas.  The total
amount of carbon accumulating annually together with sediments in the epicontinental zone of the ocean constitutes 75.7% of the total mass of carbon
introduced annually into the sediments of the World Ocean.  The shelf region
accounts for as much as 75% of the overall amount of organic carbon; the
percentage of organic carbon in the total mass of carbon increases toward
the coastal zone and attains a maximum in sediments of inland water bodies.
The mass ratio of organic carbon to total carbon in various sediments
amounts to (in %): deep ocean, 12.3; continental slope, 34.2; shelf, 40.4;
lakes, 78.8; swamps, 100.0 (Uspenskii, 1970).

Consequently, the production of hydrocarbons and the formation of their
deposits are primarily associated with the zone of transition from sea to
dry land, where the greatest amounts of organic matter accumulate and the
most favorable conditions for its burial exist.  The overwhelming majority
of fossil fuel deposits are thus distributed in sediments which accumulated
on vast territories on both sides of the ancient shorelines.


## 3.  RELATIONS BETWEEN DRY LAND AND SEA

The main link between dry land and sea is via rivers or wind action.
The wind introduces into the seas and oceans dust, volcanic ash and, less
frequently, larger pelitic or silty material, and removes huge masses of
water vapor.  Rivers transport, in suspension or solution, the bulk of the
material resulting from the weathering or leaching of rocks.  They carry
into the seas chemical elements typical of the dry land, complete the hydro-
logic cycle and, together with living organisms, take an active part in the
geochemical cycle of numerous elements.

According to Uspenskii's (1970) rough estimates, there are about four
million rivers on the earth, the combined length of their drainage systems
exceeding 40 million km.  Their total volume of water at any moment is
about 1200 km$^3$.  Rivers account for only 0.01% of the various types of surface water, but their annual throughflow is 30 times greater than their combined volume at any one moment.

In the words of A. I. Voeikov, "rivers are the product of climate." River
drainage is primarily determined by the evaporation and condensation of
water, which are in turn controlled by the climatic factor.  It also depends
on other factors: the relief of the surface on which atmospheric precipitation
falls, the structural pattern and the geotectonic regime; but climate is the
dominant factor.

The rivers on the earth's surface incise extensive channels: river beds
and valleys.  Rivers always flow toward the lower-lying or depressed elements of the relief, contributing greatly to its dissection.  M. V. Lomonosov
pointed out in his time that the action of flowing water "brings about great
changes on the earth's surface." The hydrographic network is produced by
complex physiographic processes.  Its location and trend are mainly determined by tectonic factors which exert a considerable influence on the formation of river systems and their location; many river networks are characterized by fairly ancient, inherited patterns.

Rivers, being one of the stable topographic features, have a rather considerable effect on the distribution of terrigenous and even carbonate sediments.  Kuenen's calculations (1957) showed that during the present epoch
$12 km^3$, i. e., more than 99% of the total amount $(12.12 km^2)$ of clastic material
introduced into marine basins annually, is removed from land areas by
rivers, and less than 1% accumulates as the result of abrasion.  Rivers have
always played a key role in the evolution of the earth's sedimentary cover,
serving as a connecting link between dry land and sea.

Many geologists are of the opinion that tectonics is the sole factor determining the thickness of sedimentary formations.  Strakhov (1962) rightly
claims that this viewpoint is oversimplified and that the actual mechanism
governing the thickness of sedimentary deposits is much more complex, and
that the factors involved in this process are much more varied.

According to Strakhov, sedimentation in its most general aspect amounts
to the following: mobilization of the source rock material through weathering
or some other mechanism — transport of the sedimentary material and its
partial deposition along the transportation path — introduction of the remaining material into terminal water bodies and its ultimate deposition there,
depending on the physicochemical and hydrobiological characteristics of the
basins — lithification of the sediments.

Strakhov and other authors have discussed various stages of lithogenesis
of the humid type, corresponding to the most favorable conditions for the
formation of mineral fuels.  The distribution of the products transported by
rivers from catchment areas is of particular interest from the viewpoint of
our studies.  The suspended material carried by rivers is partly deposited
within the river bed and on the floodplain, forming alluvium.  Most of the
dissolved material is carried into the terminal water bodies.

Alluvial deposits are very common among continental deposits of all geologic ages.  They are formed at all stages of river valley development.  The
only variables are the concomitant erosion and accumulation, which are dependent on the climate, tectonic regime, topography, the hydrodynamics of
water flow and other factors.  The term "alluvium" refers to fluviatile,
floodplain, and oxbow-lake deposits, as well as intradeltas forming vast alluvial plains generally filled with fluviatile deposits of all three types.  Such
accretion plains are usually formed along the lowlying coasts of marine

basins.  The underwater portions of deltas (prodeltas), bars, spits, barrier beaches and other coastal features, such as beaches and dunes, are formed under marine conditions or under the predominant action of waves and winds. They are generally not considered as alluvium, even though they are similar to it in some respects.

Detailed studies of alluvial deposits are available in the works of N. I. Nikolaev, E. V. Shantser, Yu. A. Zhemchuzhnikov, G. I. Goretskii, L. N. Botvinkina, P. P. Timofeev, A. P. Feofilova and V. S. Yablokov.  The detection of ancient alluvium, in general, and the establishment of the features of the ancient river beds and paleodeltas, in particular, are of great theoretical and practical importance.

Alluvium is very heterogeneous lithogenetically.  It contains various types of sediments which are characterized by frequent changes in conditions of deposition, structural and textural features, variations in the mode of deposition, etc.

Inhomogeneity is also characteristic of other, particularly, nearshore-marine, deposits.  Therefore, the lithologic composition alone is insufficient for assigning a deposit to a particular genetic type.  However, as pointed out by Shantser (1966), true complexes of sediments and rocks referred to the same genetic type are never random combinations of miscellaneous lithogenetic types.  They always are ordered parageneses whose interrelationship is expressed by common structural regularities.  They all have characteristic features produced by identical or similar conditions of accumulation, and show the same type of alternation and change in section and along the strike.  Consequently, knowledge of the laws governing paragenetic combinations of deposits of different lithologic composition specific to a given genetic type is one of the basic prerequisites for determining the environments in which each of them originated.

According to Shantser, the term "genetic type" does not refer to the type of sediment, but rather to the type of deposit, i. e., the category uniting complexes of sediments and sedimentary formations which are related to one another in terms of their structure and history of formation.  The identification of genetic types is necessary for determining the environments of deposition of the constituent sedimentary formations.  Paleogeographic analysis serves the same purpose in that it aids one to reconstruct the physiographic environment of deposition of particular strata and to detect the regularities of their variation in time and space.

It is not our intention to define such concepts as genetic types of deposits or facies.  They have been given sufficient coverage in numerous specialized works dealing with the process of sedimentation.  However, since we shall make use of these concepts, it is necessary to point out that different authors treat them differently.  There is less disagreement concerning genetic types. This term generally implies a complex of rocks characterized by similar genetic features.  Two major trends are discernible in the interpretation of the term "facies," which is used very loosely.  In some cases it refers to the characteristics of sediments, in others to the conditions of their deposition.

We shall not dwell on the term "facies," but prefer to point out that in practice the concept of the facies of a particular sediment involves, first of all, the determination of its lithologic composition, texture, paleontological remains, the character of bedding and transition to other varieties along the strike, and then reconstruction of the environments of deposition.  It is self-evident that the environments in which the deposits originated and consequently their facies classification cannot be ascertained unless their composition and properties are known.

But this is not all.  Facies analysis can only be of help in reconstructing the physiographic sedimentary environment if it reveals the regularities of spatial variation and association of different facies.  Facies analysis used for the purposes of paleogeographic reconstruction must have a certain specificity determined by its goals.  In this context, some of Shantser's (1966) conclusions are of particular interest.

He rightly claims that in most cases it is impossible to adopt a correct approach to the investigation of a given facies or to determine its true paleogeographic features, if its connections with the neighboring facies are ignored and if its position within the entire sequence of deposits constituting a particular stratigraphic horizon is not well established.  The size, shape, position of boundaries and qualitative characteristics of the facies reflect the changes taking place during the process of sedimentation, while the facies proper is kind of a derivative of these variations.  The identification of facies as the main technique in facies analysis is compared by Shantser to the method of differentiation in mathematical analysis.  Depending on the purposes of the study, identification of facies may be based on various characteristics or combinations of characteristics of rocks, which serve as different "independent variables."

Large facies divisions with distinct characteristics may have an independent significance in reconstructing the general features of ancient physiographic environments.  In order to reconstruct these environments and the course of sedimentation in greater detail, finer subdivisions are necessary, which in this case acquire a wholly independent significance and may be viewed as separate facies, while larger sequences are considered as groups of facies.

In Shantser's opinion, the facies is "not so much an object of study, calling for accurate, formal determination, but rather a means of investigation; thus, the specific meaning attached to this term is largely dependent on the particular problem to be solved in each particular case."  Proceeding from the above definition, one may conclude that the use of facies analysis in paleogeographic reconstructions is one of the main procedures aiding one to determine the physiographic conditions of accumulation of certain strata or horizons.  Moreover, one should bear in mind that paleogeographic investigation is not an end in itself, but should rather be regarded as an important additional source of information on the distribution patterns of producing formations and as a guide in selecting an optimal trend in exploration work.

## Coastal Zone

We regard this zone as including the coastal part of the dry land and the outer margins of water bodies.  In fact, it is a transitional zone between the lithosphere and the hydrosphere.  It is characterized by the most intensive interaction between these two spheres themselves, on the one hand, and

between them and the biosphere and atmosphere, on the other.  It is here
that the most favorable environments are created for the formation and
burial of organic matter.  Such zones stretch not only along seacoasts,
but also along the shores of other large inland basins.

We consider the term "coastal zone" to have a looser meaning than the
term "littoral zone" used by oceanographers to signify a region which is
developing under the influence of a common factor — wave action (Zenkovich, 1962).  The littoral zone, according to Zenkovich, consists of the following three elements: 1) the sublittoral — the shallow-water part of the sea
bottom whose relief is produced by waves at a given sea level (the water
overlying the mentioned sublittoral is termed the neritic); 2) the supralittoral — the strip of land with relief forms produced by the sea at a given
mean level (barrier beaches, offshore bars, or other accretion forms); 3) the
intertidal zone — the zone of activity of breakers, its width depending on the
surface slope of the detrital deposits or bedrock and on the wave parameters
(surf sometimes gives rise to beaches which occupy part of the supralittoral
and part of the sublittoral).

Such a subdivision, though justified for the purposes of modern oceanography, is virtually unfeasible in paleogeographic reconstructions because
of the absence of traces of the ancient shorelines or even more stable topographic features.  We are, therefore, perfectly justified in using the term
"coastal zone" to convey a certain mean position of the transitional region
between dry land and a marine basin during a period in which the position
of the shoreline is more or less stable.  In other words, we are dealing
with a broader stretch of the continental and nearshore marine regions than
that recognized by the oceanographers.  In any event, the paleogeographic
coastal zone is in principle not contrasted with the present-day shore zone,
but is, on the contrary, compared with it.

Numerous studies of present-day sea shores and floors have yielded
much data pertinent to comparative paleogeographic investigation of the
zone of active interaction between the earth's shells or spheres, primarily,
the land and sea.  Various aspects of this interaction are being studied for
different purposes.  Certain results obtained by such branches of science
as the study of seashore development, the oceanography of coastal waters,
hydrogeology and certain others, may make an important contribution to
paleogeography and petroleum geology.

Many Soviet and non-Soviet scientists have published comprehensive
studies on the development of present-day shore zones.  Zenkovich's (1962)
fundamental monograph serves as an example.  We know much less about
ancient coastal zones and their formation, although the most valuable mineral
resources are distributed within them.

The general features of ancient shore zones were apparently identical to
those of their present-day counterparts; both are characterized by specific
relief forms and sedimentation processes.  Whereas the geotectonic and climatic conditions have changed and the topography of the earth's surface has
taken on new forms, the principal laws of hydrodynamics and aerodynamics
remain the same.  Thus, in those places where land and sea came into contact, particular types of sediments were always deposited.

It is quite evident that ancient coastal zones which have lost many of their
original features will differ in appearance from present-day ones.  One

should not, therefore, expect to be able to detect them unless subsequent superimposed changes due to endogene and exogene factors are taken into account.

Numerous works by Soviet and non-Soviet authors have dealt with various aspects and features of ancient alluvial and nearshore-marine formations. Although almost all of these aspects and peculiarities have been established from Recent deposits and natural rock outcrops, some problems and controversial issues remain. Reconstruction of past physiographic environments is a most difficult task, especially if the petroleum geologist has at his disposal only very limited core material and well logs.

In such cases one has to resort to deduction and reconstruct the paleogeography on logical grounds using the few data at hand. One should not be apprehensive about deductive constructions and diagrams which later can be checked and refined for the plotting of more detailed maps. The comparative method and the regional approach to evaluation and interpretation of the available material are essential prerequisites for proper construction. In practice, it is virtually impossible to determine accurately the depositional environment of a certain horizon or producing formation without also investigating its relationship to the adjacent strata or without utilizing the results of studies of Recent sedimentogenesis. Even if the most detailed structural map of an oil- and gas-bearing territory is available, one cannot do without the above information.

*Accretion Forms of the Coastal Zone*

Sedimentation in coastal zones of marine basins is generally controlled by the movement of water, i. e., by currents. Zenkovich (1962) distinguishes two principal groups of currents. The first includes currents which manifest themselves simultaneously with waves and are due to the same cause. They are an active factor in the displacement of the sand deposits. The second group consists of currents which are not associated with waves and are due to other causes. This group is rather varied. It includes both drift currents and currents caused by the general circulation of waters in the basin in question, resulting from differing densities of the water, river runoff and the conditions of water exchange with neighboring basins. It also includes temporary currents caused by changes in atmospheric pressure or other factors. The velocity of these currents decreases with depth and their direction changes at a distance from the coast.

Zenkovich emphasizes the fact that near the shore permanent currents rarely attain an intensity enabling them to move the bottom material. They may exert a considerable influence on the bottom only in the relatively deep-lying portions of the sublittoral, but even there their action is much weaker than that of wave currents. Permanent currents generally have a maximum influence on the bottom beyond the sublittoral.

Wave motions normal to the shoreline are responsible for such accretion forms as barrier beaches which are very common in the coastal zone. Zenkovich and Leont'ev estimate that modern barrier beaches and lagoon beaches account for about 13% of the overall length of the earth's shorelines. Many bars attain widths of several kilometers.

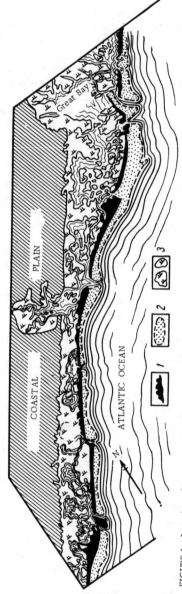

FIGURE 1. Location of bars on the Atlantic coast of the U.S.A. (W.A. Ver Wiebe, 1959):

1 — above-water bars;  2 — subaqueous bars;  3 — swamps and marshes.

Drilling operations in the Gulf of Mexico have revealed a bar with adjacent lagoon deposits. The base of this bar is at a depth of more than 10 m, having the form of a prism sloping slightly toward the shore. The bar borders on marine sands on the outer side, while on the inner side it is contiguous with various types of lagoon deposits. The absolute age of the bar's lowermost portion, according to radiocarbon dating, is 5000 years. The shore has remained essentially in the same place throughout this entire period, shifting landward and seaward within some 3 km. The waves had sufficient time to deposit the enormous body of sediments forming the bar. The sand material was brought to the Gulf by rivers, and was subsequently spread along the sublittoral and the supralittoral.

The present-day barrier beaches on the Atlantic coast of the U.S.A. have the form of sand strips up to 10—16 km in width. In places the bars are cut by tidal scours. Shallow-water lagoons and marshes (Figure 1) are located on the nearshore side of the bars. Some of the bars extend for hundreds of kilometers. Other huge bars are known from the coast of the Great Australian Bight. The Arabat Tongue, separating the Sea of Azov from the Sivash Lagoon, has a length of about 200 km. The Meechkyn spit on the Chukchi coast is more than 100 km long. The common feature of all these bars is that they consist of bottom sediments.

The British geographer King (1963) has described a barrier beach on the southern coast of Iceland, which extends for 215 km from Vik Cape in the west to Hornafjördur in the east. The largest bar is near Ingolfshofoi Cape. It is separated from the shore by a shallow-water lagoon, 6 km wide and 0.3 to 0.6 m deep. In places the lagoon narrows considerably, and fluvioglacial deposits come into contact with the bar. The erosibility of the tuffaceous rocks of this region, in conjunction with the active removal of eroded material by glacial waters, ensures an abundant supply of sediments for the formation of the beaches.

When lacustrine basins break through ice sheets, an enormous amount of material of varying mechanical composition is carried with the water onto the shores; sand is the most important component for the building of barrier beaches. Two other bars composed of sand material are situated to the west of the above-mentioned bar. The median diameter of the sand particles making up the outer bar is 0.42 mm, and that of the inner bar, 0.39 mm.

According to King, a necessary condition for the building of a barrier beach of this type is the presence of inwash waves, and, accordingly, of an appropriate wave and wind regime. Since the sand constituting the barrier beaches of Iceland is rather coarse, the inwash waves on these beaches must be steeper than those producing fine-sand beaches. A similar development of barrier beaches is, according to the mentioned author, also characteristic of other coasts where the tide is not high and the coastal zone has a gentle slope. This applies to the Gulf of Mexico, the Baltic Sea, the Mediterranean and the greater part of the Atlantic coast of the U.S.A.

Leont'ev points out that King has totally disregarded the fluctuations in the level of the World Ocean; nor does he emphasize the importance of such preconditions as the gentle slope of the original sublittoral and the presence of loose material on it. We are paying so much attention to the subject of bars because they are such common aggradation features along seashores, both present-day and ancient.

In addition to barrier beaches, there are also submarine bars which are generally rather narrow aggradation ridges of asymmetric structure, the steeper slope facing the coast.   The Bakal, Odessa and Yevpatoria banks in the Black Sea are bars of this type.   Submarine bars have been discovered in the Caspian and other seas.

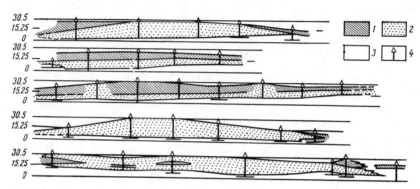

FIGURE 2.   Cross sections of sandbars (Ver Wiebe, 1959):

1 — sandy shale;  2 — sand;  3 — shale;  4 — wells.

The paleogeographer is primarily interested in ancient bars.   Ver Wiebe (1959) cites examples of ancient barrier beaches detected during drilling for oil and gas in east Kansas, in the Greenwood and Butler regions.   Petroleum occurs there in extensive narrow strips in Pennsylvanian (Late Carboniferous) formations.   These "shoestring" sands have a flat base and a somewhat concave surface (Figure 2).   They are productive in those places where their thickness is 15 to 30 m.   Individual pools are 3.2 to 9.6 km long and form strips 40 to 72 km in length.   Their average width is about 1.6 km.

As these sand formations extend along a shoreline, they should not rightly be called "shoestring," the latter term implying primarily sandstones produced in river channels or by submarine currents.

Other accretion forms of sand formations are also encountered in the coastal zone, in particular, dunes, spits, and barrier beaches, which are not easy to detect in the fossil state, but have certain specific features.   They may be studied by way of comparison with their present-day counterparts. Dunes usually occur in localities along the coast where the necessary sand material is derived from the beach.   The conditions for this are more favorable if the sublittoral is shoaly and consists of sand.

Zenkovich believes that the factors governing the formation of dunes in a particular place are determined by the shore profile and the rate of development of the shoreline.   He thinks that eolian forms of coastal deposits develop most intensively during and some time after oscillatory movements.

The dunes in most cases form several ridges, the slopes of which become less asymmetric as one moves inland from the coast, giving rise to isolated hillocks.   The ridges or the belts of dunes are never very wide.   In the Baltic region their width does not exceed 3 to 5 km, and on the Atlantic coast of France it is about 10 km.

Dunes are classified as eolian formations whose principal features are structural asymmetry, relatively great thickness and diversity of bedding. The latter is generally of the cross or crisscross type. The cross-bedded series and the dips of the beds may show great variance. Large series, whose thickness sometimes reaches several tens of meters, predominate. The beds have a range of shapes: concave or convex, less frequently recti-linear, in places wedge-shaped, etc. Such series are often not unidirectional.

Ancient dune deposits are difficult to recognize on the basis of drilling data. They are more readily identifiable if several characteristics are em-ployed, primarily, texture, shape of the sand bodies, and their location relative to marine and continental formations. Another indication of ancient dunes is the well-sorted sand material, as well as the lagoonal or palustrine facies of the underlying rocks. Eolian sands, which are much better sorted than other types of sand, may serve as good reservoirs for oil and gas. Some American geologists (Ver Wiebe and others) maintain that individual oil pools in the Rocky Mountain region occur in dune sandstones of Jurassic age (the Nugget and Sundance sandstones).

Spits, barrier beaches and tongues, which are very common on gently sloping shores, are foci of nearshore-marine accumulation. These forms of sand accumulation extending into the sea are in most cases produced near curved, lobate shorelines or in areas adjacent to delta zones of rivers. They form relatively thick barlike bodies attaining several tens of meters in height.

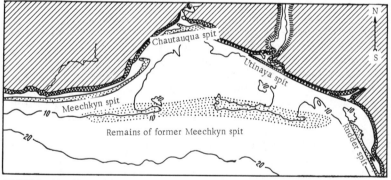

FIGURE 3.  Meechkyn spits and barrier beach in the Gulf of Anadyr (Zenkovich, 1962)

Strips of sand deposits in contact with dry land or separated from it are referred to as spits. They are not readily distinguishable from barrier beaches, especially in the fossil state. The Meechkyn bar on the shoaly coast of the Gulf of Anadyr is an example of such a present-day "spit" (Fig-ure 3). Remains of the former spit are located nearby at a depth of 10 m. The length of the marine spits on the Black Sea shore near Tendra is 66 km, and the Astrakhan spit in the Caspian Sea is more than 70 km long. Several spits-tongues extend from east to west along the northern shore of the Sea of Azov.

Apparently, every single sea or large water body on the continent has beaches, spits or other accretion forms, varying in shape and size, formed by the action of waves or the accompanying currents. Their outlines are extremely varied, but all of them are associated with the coastal zone.

According to Zenkovich, spits and tongues are narrow strips of alluvium above the water line. They may either be in direct contact with the seashore or separated from it. Spits generally include forms which run parallel to the shoreline and curve into gulfs or bays. Tongues are spits directed seaward; barrier beaches are deposits produced by wave action, which separate the mouths of gulfs or bays from the sea. All the above-mentioned accretion forms are similar to bars in general features, but they are not as extensive.

There are also other forms of coastal accumulation of the sand material. Some of them are considered below. Sand formations play a rather important role in the coastal zone. To cite an example, the bars along the shores of the Mexican Gulf extend for 2000 km. The Florida Peninsula is an enormous spit. The Western Kamchatka coast is fringed by bar deposits for nearly 600 km. The already-mentioned Odessa bank in the Black Sea is 80 km long, 10 km wide and 15 m high. Its volume is, therefore, about 12 billion m$^3$.

These sand formations generally grade into clay or silt sediments along the strike, and are buried among them when subsidence takes place. This combination of deposits is rather favorable for the formation of oil and gas pools in sand reservoirs during later lithogenesis.

The oscillatory movements and shifts of the shoreline are responsible for the frequent alternations of arenaceous and argillaceous rocks in the coastal zone. The sand deposits of this zone are characterized by their lenticular form and regional pinch-outs. Each lens may, under favorable conditions, develop into a lithologic trap containing oil and gas.

In the U. S. A. pools associated with pinch-outs of sand reservoirs are known in the Appalachian oil and gas basin, in the south of Texas, along the Gulf Coast, in the Western and Eastern Interior basins, and in many other localities.

Oil and gas deposits occurring in ancient buried bars, spits and barrier beaches are especially widespread in the eastern part of the Western Interior basin; they are particularly numerous in Pennsylvanian sand horizons. The lithologic factor is the controlling one there.

The richest pools in this basin are associated with ancient bar deposits truncated by anticlinal folds and also with large monoclinal sand lenses. These are typical pools of the lithologic type. The largest of them were discovered in the oil- and gas-bearing regions of Kansas and Oklahoma.

Ancient sand bars of Pennsylvanian age are analogous to the present-day bars of the Atlantic coastal plane (Figure 4). They formed a zone of sand lenses which extended along the ancient Cherokee Sea during the Pennsylvanian. Oil pools of the lithologic type are most characteristic of lenses of the Bartlesville and Burbank sands (Pennsylvanian). The Burbank field is one of the largest oil deposits in Oklahoma. It is located on the western slope of the Ozark uplift, where the Carboniferous strata dip westward at an angle of less than 1%. The producing sandstones, in the form of elongated lenses, are in fact buried offshore bars whose thickness ranges from 7 to 30 m. Pools associated with lenses of Bartlesville sand in the Glenn field, for instance, are of the same lithologic type.

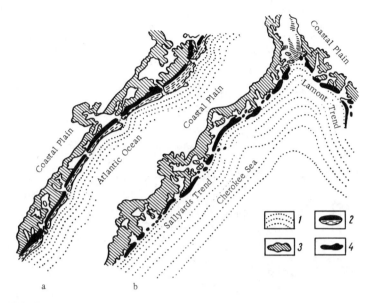

FIGURE 4. Comparison between ancient and present-day offshore bars (Rich, 1938):

a — part of the present coast; b — ancient shoreline during Cherokee times in southeast Kansas;
1 — seawater area; 2 — present-day bars; 3 — lagoons; 4 — lenses of Bartlesville sand.

Russel (1958) gives examples of petroleum occurring in sand lenses of the bar type. These are Devonian lenticular sandstones in the northern part of the Appalachian oil and gas basin, the Davis sand lens of the Hardin field in Texas, the Richburg and other pools in the state of New York, and the Mirando oil and gas area in Texas.

The oil in the latter is produced from lenticular sand bodies of Eocene and Oligocene age; 15 pools in this region are of the lithologic type. The deposits there have very distinct lithologic features. Most of the pools in this region run parallel to the present and ancient shorelines. All of them can be classified as buried bar formations.

Pools of the stratigraphic type are also widespread in the coastal zone, alongside structural and lithologic pools. Such deposits usually occur in monoclinally dipping sand-silt beds truncated by erosion and unconformably overlain by impermeable rocks. The pools of the Oklahoma City field are of this type; the Ordovician sandstones there are unconformably overlain and sealed by Pennsylvanian rocks. Stratigraphic pools have also been found in some regions of Central Kansas, both on the slopes of large uplifts and on the flanks of domes or local positive structures.

Much factual material on petroleum occurrences in the U. S. A. and other countries is available in the studies of Levorsen (1954—1965) and Bakirov (1959), in the reference work "Geologiya nefti" ("Petroleum Geology") edited by Vysotskii (1968) and Vasil'ev (1969), and in other sources. The monographs by Maslov (1968) and Gostintsev and Grosgeim merit particular mention as they deal specifically with lithologic and stratigraphic pools; they cover aspects of the distribution of and search for oil pools of this type.

We could cite numerous examples of oil and gas fields and pools of all types associated with coastal formations of various ages.  As we have repeatedly emphasized, this is mainly due to the fact that the sediments in which the oil is generated and those in which it later accumulated were deposited either next to or in close proximity to one another.  However, environments favorable for oil and gas generation and accumulation could have also arisen far from the coastal zone, in sediments of extensive shallow seas as well as in freshwater basins of the remote past.

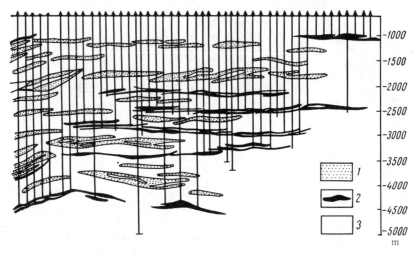

FIGURE 5.  Cross section of the Goose Creek oil field (Russel, 1958):

1 — sandstones;  2 — oil-bearing sandstones;  3 — clay.

We shall refer to the cross section of the well-known Goose Creek oil field in the U.S.A. (Figure 5), where lithologic pools are located in Miocene sand lenses.  The shape of the lenses suggests that they are, in all probability, small bars and spits formed on a shoaly beach.  They are interbedded with argillaceous, apparently oil-producing rocks, and the entire series occurs at a depth interval of 300 to 1400 meters.  In the Goose Creek field, South Texas, the Paleogene-Neogene sand series were formed under conditions of the gradual retreat of the waters of the Mexican Gulf.  This regression was accompanied by brief periods of transgression and ingression. The pinch-out zones of the sandstone in this region run parallel to the Gulf Coast.

## Embayed Coasts

In studying any type of accretion form of the coastal zone one should bear in mind that the shoreline is not only unstable but also that it seldom remains straight over large distances.  In most cases, it is curved and in places there are bays and capes of various size and shape. Embayed coasts are not infrequently the result of inundation of the land edge by the sea; in most cases erosion or tectonic depressions become flooded. Such coasts are of the ingression type and may show great diversity.

The sea dissects the coast due to the unequal resistance of the rocks on the relatively smooth edges of the subsiding land. The coast is usually more markedly dissected in cases when lowlands are inundated. Zenkovich distinguishes uneven (open and closed), lobate and other types of embayed coasts.

During the development of an embayed coast there is close interaction between abrasion and accretion processes which are essentially dependent on the vertical tectonic movement of the coast. The morphology and structure of the coast are very important factors. There is a sequence of sections on embayed coasts: those from which clastic material is removed, areas where this material accumulates, and intermediate areas through which it is transported.

According to Zenkovich, under natural conditions the rate of development and configuration of an embayed coast, determined by the size and location of the accretion forms, are also dependent on the type and amount of detrital material brought to the coast during the abrasion processes and by runoff, as well as on the amount of loose material that covered the land prior to its submergence.

This accounts for the great variety of combinations of forms of embayed coasts. The nature of the land, whose submergence is usually responsible for the shaping of an embayed coast, contributes in equal measure to the variety. Significant roles are played by the relief-producing processes which were active prior to the inundation of the territory, the geological structure of the territory and, finally, the depth of the subsequent subsidence. Zenkovich thus arrives at the conclusion that there cannot be any universal scheme accounting for the development of an embayed shore. The researcher, in his analysis of the modern forms, is therefore obliged to study the effect of each such process and possess knowledge of the factors determining it. This is the only way to identify the principal process in each area and to carry out an analysis of the evolution of the entire variety of accretion forms.

The above work procedure is recommended to paleogeographers studying buried features of ancient coastal zones. There are equal chances of discovering oil and gas in sand reservoirs of embayed coasts as in bars, spits and similar formations.

Oceanographers specifically distinguish another type of dissected shore: lobate. Its principal feature is the irregular configuration of individual water areas as well as of the separating peninsulas, capes and islands. This type of shore comes about when lowland territories with a complex relief of tectonic character are flooded by the sea. This results in the formation of rather large, variously oriented water areas within which the wave action is totally independent of the open sea. Zenkovich gives examples of present-day lobate shores showing various degrees of dissection: several interconnected water areas of the Chernaya Inlet on the southern island of Novaya Zemlya, as well as a large portion of the Black Sea shore between Western Crimea and the Dnieper liman. Shore development there is due to the different internal systems of waves in each bay, rather than to the waves of the open sea.

This weak wave activity may cause abrasion and produce accretion forms only if the rocks are poorly resistant. The generally rounded outlines of the abrasion areas, together with the accretion forms, indicate that the shore is generally very sensitive to changes in the sediment load of the streams. The irregularity of the lobate shore between the Crimea and the Dnieper was further increased by a number of long spits separating extensive water areas from the sea.

Embayed shores fringed by bars are rather common on the Atlantic coast of the U.S.A. and on the western shore of the Sea of Azov where a huge bar — the Arabat Tongue — separates the highly dissected inner shore of Sivash from the sea. Zenkovich relies on the following features in reconstructing the earlier phases and general course of development of present-day, gently sloping embayed shores:

1. Marked predominance of zones of accretion over abrasion areas along considerable stretches indicates that most of the detrital material was derived from the bottom and that the shore was initially shoaly.

2. The character of the material constituting the accretion forms is often a direct indication of its origin. For instance, if the barrier beaches are made up of sea shells, then abrasion could not have been involved in their formation. The above-mentioned Arabat Tongue, in particular, consists of sea shells.

3. The structure and bottom relief, as well as certain lithological features of the bottom deposits, are important indicators.

It is quite evident that the laws governing shore development in past geological epochs were essentially the same as those applying at present. Consequently, the mentioned characteristics may be helpful in reconstructing ancient shores.

*Beaches*

Ancient beach formations are not easy to detect. However, like other coastal accretion forms, they may serve as reservoirs in the land-to-sea transition zone. Shepard (1969) classifies modern beaches according to a number of features. The beach profile is one such characteristic. It may either be gently sloping or have a slope complicated by one or more beach berms. Sometimes a subaqueous accretion terrace is formed, either with or without bars and rills.

Greater differences between beaches may be due to the grain size of the constituent materials: gravel, pebbles, coarse or fine sand. A typical gravel beach has an offshore bar whose height may attain 5 or 6 meters. Some bars are composed of shells.

If there is a berm on a coarse sand beach, it usually slopes toward the land, sometimes at a considerable angle. The beach slope is dependent mainly on the permeability of the constituent sand material. The coarser the grain of the material, the steeper the frontal slope. A very gentle frontal slope is characteristic of fine sand beaches.

Beaches also differ according to the tide range. Those on tideless seas are often fringed by a number of offshore bars and rills, whereas those influenced by high tides are characterized by broad terraces, each usually accompanied by a single large bar. Crescent-shaped bars, concave toward the land, are formed in some tideless seas.

The character of the beach is to a large extent determined by the configuration of the shore. Long and rather straight beaches of the barrier type may form between widely spaced promontories. Short, crescent-shaped or "bay" beaches are generally squeezed between steep headlands jutting into the sea. In most cases they consist of coarse-grained material derived from disintegration of the cliffs fringing the bay, whereas fine sand material brought by the waves is more typical of long beaches.

Sediments for building up beaches are derived mainly from the adjacent areas of the sea. Nearly all the sand material required is brought from the sea bottom. In tropical regions much sand is produced by the disintegration of reefs and shells of various organisms. River sand transported over great distances by littoral currents is apparently another source of supply. This is partially confirmed by the fact that in places where rivers flow into deeply incised bays (estuaries), the coastline between these estuaries often lacks beaches. Apparently, this is the result of the deposition of nearly all of the sand material brought by rivers near the river mouths.

According to Shepard, beach sands are quite diverse in their mineral composition. Quartz is the most common component in most sand beaches of today. Feldspar, mica and other minerals are much less abundant. Magnetite and ilmenite accumulations are often encountered at the base of sand deposits in the rear zone of the beach.

The waves not only cause differentiation of the terrigenous material, but are also responsible for the bedding of the beach deposits. The bedding shows up most prominently in the frontal slope, where it is either parallel to the slope surface or oblique to it, having a very gentle dip. The bedding is accentuated by admixtures of dark minerals (hornblende, magnetite, etc.) or accumulations of shells and their detritus. The boundaries of sets of beds may have irregular shapes.

According to Botvinkina (1965), beach deposits, as distinct from bar deposits, are not characterized by the presence of layers differing in texture within the cross-bedded deposits. The regular pattern of the bedding of the beach sediments, especially common on the shores of bays, lagoons or gulfs, is often disturbed by "air pockets" which cause contortion of the beds, leading to their vertical orientation.

Beach deposits often border on lagoonal ones or are covered by the latter during marine transgressions. During regressions they may be covered by eolian sediments. Beaches thus may become buried both during transgressions and regressions; in both cases they will be situated at the boundary between marine and continental deposits. One should, however, recall that sediments of marine shoals may show bedding similar to that of beaches. The former can be distinguished by the fact that they are surrounded by marine deposits on all sides.

There are hardly any references in the geological literature to the occurrence of oil and gas in ancient beach deposits. This should probably be attributed to their great similarity to other sand accretion forms, rather than to the fact that they cannot serve as reservoirs. Moreover, beach formations can be mistaken for offshore spits or bars, if judgment is based solely on borehole data.

We have mentioned beaches so that the geologist, in his reconstruction of paleogeographic environments in the transition zone from continental to marine facies, will also focus his attention on these potential reservoirs. In such cases, the sandstones and siltstones have a sheetlike distribution, the thickness of the sheets usually decreasing from the coastal zone toward the sea. Strong tidal currents or frequent nearshore swells may also give rise to the formation of blanket sand deposits.

### Lagoons

We should like to mention yet another form of accumulation of sand deposits in separated portions of the sea — lagoons. These may be of quite different types, including alluvial, insular, deltaic, depression, etc. Alluvial lagoons are the most common and the largest. For example, Laguna Madre extends for 400 km along the Gulf Coast, from the delta of the Brazos River to the delta of the Rio Grande. Beyond the small delta of the Rio Grande, the lagoon continues south for a further 300 km. Its width ranges from 20 to 40 km, attaining 50 to 60 km in the limans of river mouths.

Figure 6 shows the distribution of sediments in the Kurskii Zaliv (Courland Lagoon) on the south coast of the Baltic Sea, at the mouth of the Neman River. This lagoon, which is of the alluvial type, extends for 93 km along the shore; its maximum width is 44 km. Fine sands fringing the shores are the predominant sediments there; they are especially abundant in the numerous channels of the Neman delta.

Deltaic lagoons, which are formed as a result of the separation of large bodies of water from the sea by spits and sand necks, are also quite common. The formation of such lagoons is promoted by sand brought by rivers, which is deposited by waves and littoral currents to form separating spits. In the northern part of the Mississippi delta there is a lagoon — the Pontchartrain Lake — which is 70 km long and 35 km wide. Another, somewhat smaller lagoon — Borgne — is situated further to the southeast. Yet another lagoon — Barataria — measuring 30 by 20 km and with a depth ranging from 1.5 to 2 m, is situated on the western edge of the Mississippi delta. The sediments of this lagoon are arranged similarly to those of the Courland Lagoon. Shore and surf-zone sands lie along the margins of the lagoon. Toward the center they give way to sands with a small admixture of clay particles, then to argillaceous sands in which sand material accounts for up to 25%, and, finally, to clays with up to 15% of sandy material.

As pointed out by Nalivkin (1956), the sand admixture is often due to the action of winds which carry sand from the spits and shores into the lagoon. Violent storms, especially hurricanes, effect great changes in the distribution of sediments in the mud facies. Unfortunately, this is seldom taken into account in studies of sedimentary environments. There are several known examples of petroleum occurrences in deposits of ancient lagoons, especially those of the deltaic type. These include the giant Athabaska oil field in Western Canada, which is an enormous sand lens. The Lower Cretaceous sands of the McMurray formation are impregnated with inspissated oil and

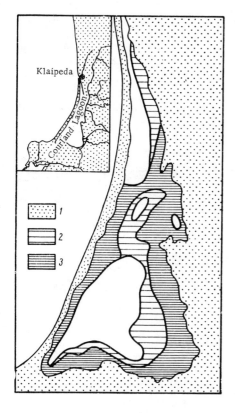

FIGURE 6.  Distribution of sediments in the Courland Lagoon (Nalivkin, 1956):

1 — sand content in the mud ranges from 0 to 35%; 2 — sand content in the sediments ranges from 35 to 65%; 3 — sand content in sediments ranges from 65 to 100%.

separated from the underlying Devonian carbonate rocks by an erosion un-conformity.  The sands are represented by alluvial-lacustrine and fluviatile facies and are composed mainly of coarse-grained quartz material.  They cover a vast territory and range in thickness from several centimeters to 90—100 meters.  The sands pinch out toward the source area and are re-placed by clays in the opposite direction.

Producing sands in the present-day valley of the Athabaska River can be traced over 160 km.  Over the greater part of this territory the sands lie close to the surface and are incised by modern river valleys.  The struc-ture of sands ranges from finely bedded to massive.  Locally they alter-nate with clays and are overlain by the latter.  The sandstones pinch out on approaching elevated features of the ancient relief.  The deposition of these sediments was the outcome of the rather vigorous action of rivers during Cretaceous times.  Ancient valleys with steep walls in the limestones underlying the sand formations serve as evidence of this activity.

Drilling data have revealed the erosional character of the pre-Cretaceous surface and established the composition and thickness of the sands.  They consist essentially of white quartz grains cemented by oxidized asphaltlike petroleum.  The sand beds are usually 15 to 30 m thick, but there are locali-ties where their thickness exceeds 30 m, and in the south and north of the deposit it is greater than 90 m in places.  The accumulation of the producing sand formation must have occurred in a delta or several deltas of the rivers flowing from east to west. In any event, the role of rivers was the most active one.

A deposit of similar viscous, heavy petroleum (asphalt) — Peace River — was discovered 200 km west of the Athabaska field. Here Cretaceous sands unconformably overlie Jurassic, Permian-Pennsylvanian, and Mississippian formations. Between these two fields there are several smaller occurrences of oxidized oil. The latitudinal profile across the Athabaska-Peace River oil deposits is shown in Figure 7.

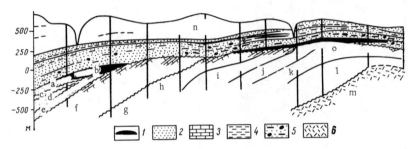

FIGURE 7. Diagrammatic profile of deposits in the oil-bearing sands of the McMurray formation (Lindtrop et al., 1970):

1 — oil pools; 2 — argillaceous rocks; 3 — limestones; 4 — shales; 5 — coal inclusions; 6 — granite. Key to inserts in figure: a) Bluesky-Getty; b) Peace River; c) Jurassic; d) Triassic; e) Permian-Pennsylvanian; f) Mississippian; g) Devonian; h) Wabamun; i) Winterburn; j) Woodbend shale; k) Beaverhill Lake; l) Elk Point; m) Precambrian; n) Glacial deposits and undivided chalk; o) Athabaska oil-bearing sands

Opinions differ as to the mode of formation of the unique Athabaska accumulation containing about 60 billion tons of oil. A.MacLeary, for instance, is of the opinion that the clays covering the McMurray sands were the source beds; G. Hume links the formation of the oil with Early Cretaceous argillaceous formations syngenetic with the sands but occurring in the deepest parts of the basin. W. Gussow voiced the view that this deposit is stratigraphically sealed; the oil penetrated into it up-dip through the highly permeable basal sands, from zones of high pressure and temperature into zones of lower pressure and temperature. The McMurray sands were the terminal traps in the path of oil migration along the unconformity surface from the huge source basin whose area was 650,000 km$^2$. The oil became heavy (0.94 to 1.0) due to oxidation. According to Th. A. Link's hypothesis, the oil migrated into the Athabaska deposit from the underlying Devonian carbonate rocks along a system of fissures; after rising almost to the surface it penetrated into the bottom sediments of barred basins along the shore, which at that time were being filled with sands of eolian and deltaic origin. There are also other points of view on the formation of this unique deposit. K. Corbett believes that the oil was inspissated by humic acids supplied to the lagoons by continental rivers.

In addition to the oil sources, the environments of accumulation of huge masses of sand material in the lagoons and adjacent areas are also of great interest. According to the latest data, the producing sands of the Athabaska field filling ancient lagoons should be regarded as deltaic and nearshore-marine formations.

We shall cite another example confirming the favorable combination of conditions for both the formation and accumulation of oil in lagoon deposits. In the Early Jurassic basin of northern Louisiana and southern Arkansas the Cotton Valley formation is productive.   The following sedimentary facies are identified: coastal plain (red sandstones and conglomerates); the Proto-Mississippian deltaic (sands); lagoon (red and dark clays and sandstones); barrier spit (massive sandstones); open-sea (clays with intercalations of argillaceous limestones).  Veber (1966) has plotted the oil and gas deposits discovered in the Cotton Valley formation on a paleogeographic map (Figure 8).  Their correlation with the sediments of a lagoon, bounded by a barrier spit in the south and by a delta in the east, is quite evident.  The productivity of the deposits increases from north to south as the sand strata become more continuous.

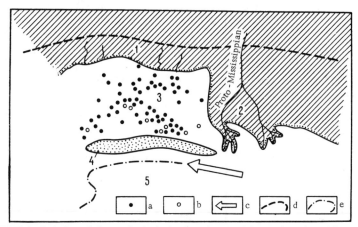

FIGURE 8.  Early Jurassic lagoon in the basin of northern Louisiana and southern Arkansas (according to Thoms and Mann with additions by Veber, 1966):

a — deposits of oil and gas condensate;  b — gas deposits;  c — direction of littoral currents; d — outlines of the pinch-out of the Cotton Valley producing formation;  e — southern boundary of the area covered by drilling;  1 — coastal plain;  2 — delta;  3 — lagoon;  4 — barrier spit;  5 — open sea.

## 4.   THE ROLE OF RIVERS IN THE DEPOSITION OF SOURCE BEDS AND RESERVOIR ROCKS

Rivers have an extremely important function in the development of the earth's sedimentary mantle.   River runoff, which is determined primarily by climatic conditions, plays a particular role in the complex process of sedimentogenesis.   The data listed in Table 1 (Strakhov, 1962) give an idea of the scale of mechanical denudation effected by the largest contemporary rivers and of its dependence on the geographic location of these rivers. The close relationship between the action of rivers and the physiographic environments of their basins is apparent from Table 1.   The life of a river depends not so much on tectonic movements as on the amount and regime of atmospheric precipitation, although the tectonic factor has an undeniable effect on the direction of flow and action of the river.

TABLE 1.  Mechanical denudation in various river systems

| River | Area of basin, $10^3 km^2$ | Runoff, $km^3$ | Solid load, million tons per year | Mechanical denudation, $t/km^2$ (or $g/m^2$) |
|---|---|---|---|---|
| Amazon ............. | 7050 | 3187.5 | 1000 | 60 |
| Congo ............. | 3690 | 1350 | 86 | 18 |
| Mississippi .......... | 3248 | 590 | 500–750 | 154–230 |
| La Plata ............ | 3104 | 600 | 96.5 | 32 |
| Nile ............... | 2800 | 70 | 88 | 31 |
| Yenisei ............. | 2707 | 548 | 10.52 | 4.0 |
| Ob ................. | 2425 | 394 | 14.2 | 6.0 |
| Lena ............... | 2418 | 488 | | |
| Ganges ............. | 1730 | 960 | 1800 | 1040 |
| Amur ............. | 1843 | 346 | 52 | 28 |
| Yangtze ........... | 1175 | 690 | 275 | 234 |
| Mackenzie ......... | 1766 | 440 | 15 | 9 |
| Volga .............. | 1380 | 255 | 25.7 | 19 |
| Zambesi ........... | 1330 | 500 | 100 | 75 |
| Niger .............. | 2092 | 293 | 67 | 32 |
| Hwang Ho ........... | 980 | 126 | 630 | 640 |
| Saint Lawrence ....... | 802 | 304 | 3 | 4 |
| Orinoco ............ | 944 | 442 | 45 | 47 |
| Tigris and Euphrates ... | 1048 | 210 | 725–1000 | 690–1000 |
| Orange ............ | 1020 | 91 | 153 | 140 |
| Indus .............. | 960 | 175 | 400 | 420 |
| Yukon ............. | 855 | 185 | 88 | 103 |
| Danube ............ | 816 | 201 | 83 | 101 |
| Mekong ............ | 810 | 387 | 1000 | 1200 |
| Kolyma ............ | 644 | 120 | 4.7 | 7 |
| Colorado ............ | 590 | 20.3 | 160 | 271 |
| Columbia ........... | 772 | 187 | 36 | 47 |
| Dnieper ............. | 503 | 53 | 2.02 | 5 |
| Irrawaddy ........... | 410 | 428 | 350 | 850 |
| Don ................ | 422 | 28 | 7.75 | 18.3 |
| Indigirka ........... | 360 | 57 | 8.5 | 24 |
| Northern Dvina ....... | 411 | 111 | 5.84 | 14 |
| Pechora ............ | 327 | 129 | 6.5 | 20 |
| Neva ............... | 282 | 82 | 0.82 | 3.9 |
| Yana ............... | 318 | 31 | 3 | 10 |
| Rhine .............. | 225 | 68.5 | 4.5 | 20 |
| Po ................. | 75 | 48.7 | 18 | 240 |
| Rhone .............. | 99 | 52.7 | 31.5 | 320 |
| Garonne ........... | 85 | 21.6 | 5.8 | 70 |
| Vistula ............. | 199 | 32.0 | 2.5 | 13 |
| Rion ............... | 13.4 | 13.5 | 8.5 | 633 |
| Terek .............. | 43.7 | 11 | 26 | 600 |

About 50 to 60% of the enormous mass of clastic material transported by rivers from land areas into water bodies is deposited in deltaic zones. According to Samoilov (1952), the Yangtze River, whose annual solid load is

275 million tons, deposits nearly half of this material, in the form of sand and mud, in the delta and on the estuarine beaches.  Of the 88 million tons of solid material carried into the Nile River delta annually, 28 million tons are deposited there.  Since the 4th century, 25% of the solid load of Shatt al Arab has been deposited annually in the intradelta.  This amount is approximately doubled if the prodelta is also taken into account.  In the case of the Don River, whose solid load is 7.75 million tons, 10% remains in the floodplain, and 15% is deposited in the estuarine beach zone.  The entire solid load brought into the distributaries of the Kiliya delta (Danube), which amounts to 83 million tons, remains in the delta or on its beaches.  About 65% of the total amount of solid material carried in suspension by the Vistula River in the period from 1895 to 1929 was deposited in the delta.  The amount of suspended material that settles in the deltas of different rivers varies from 10—15 to 80—95%.

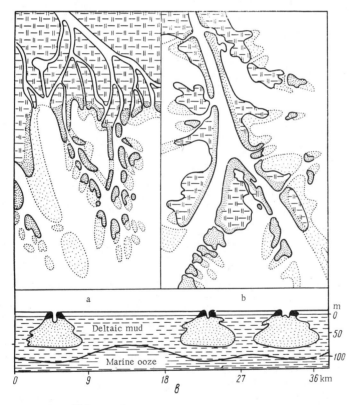

FIGURE 9.  Structure of delta margin.

Deltas:  a — Volga;  b — Mississippi;  c — cross section of the Mississippi marginal marine plain (Zenkovich, 1962).  The river distributaries are manifested as sand lenses; channel banks are shown by black shading.

The paleogeographer is interested not only in the delta, but also in the broader estuarine region which, according to Samoilov, includes part of the lower reaches of the river, the adjacent sea shore and beach (encompassing the territory where estuarine processes take place). In this region the fluvial regime is gradually replaced by a marine regime, the slope of the water surface dropping almost to zero. River runoff is the most essential and permanent factor in the development of estuarine regions; the effects of runoff are much greater in nearshore-marine formations than is presently assumed.

At its mouth a river usually flows through its own alluvium and ramifies to form arms and channels (distributaries) which are often flanked by levee-like banks (Figure 9). The type and size of the delta are mainly determined by tectonic movements in the estuarine region, the solid load of the river, the wave regime of the sea, and certain other factors. Various combinations of these factors give rise to the variety of deltaic forms.

About a hundred major rivers are discernible on modern global maps; each of them forms a large estuarine region. These regions are generally situated in subsiding nearshore areas (Samoilov, 1952; and others). We shall not go into the details of the hydrodynamic regime of the estuarine zone and the formation of its relief, but will briefly discuss the organic life there.

TABLE 2. Average composition of salts in river and sea water, %

| Basin | Chlorides | Sulfates | Carbonates | Other substances |
|---|---|---|---|---|
| River ................. | 5 | 10 | 60 | 25 |
| Sea .................. | 88.5 | 11 | 0.3 | 0.2 |

Together with detrital material, rivers also carry biogenic matter and solutions of various salts onto estuarine beaches. The mixing of the fresh water of the rivers with the saline sea water leads to new geochemical and biochemical processes (Table 2). The average composition of the salts in river waters is dependent on the nature of the relief and of the river basin. In the equatorial zone, where there is much atmospheric precipitation, river waters contain less carbonates. The high content of organic matter in these rivers causes their undersaturation with calcium carbonate. This is due to the fact that the decomposition of organic matter is accompanied by the liberation of free carbon dioxide.

Suspensions of biogenic elements may pass into solution in the delta, thus promoting the development of phytoplankton on the estuarine beach. The different solubilities of the salts and certain chemical and biochemical reactions, as well as the biogenic consumption of salts, produce, in Samoilov's opinion, the following characteristics of the hydrochemical processes on the marginal marine plain.

The main process is the dilution of the sea water by the river water, without any appreciable change in the composition of the marine salts. An equally important process — coagulation and settling of colloidal suspensions — takes place owing to the abundance of electrolytes in the sea water. As a result of hydrolysis and bacterial decomposition, the river detritus is gradually transformed into organic colloids — dissolved organic matter — and becomes fully mineralized in the course of time, with regeneration of the nutrient salts of phosphorus and nitrogen which are consumed by the aquatic microorganisms.

A considerable part of the river detritus is consumed by zooplankton while still in the stage of organic colloids. In addition, most fresh-water organisms perish in salt water and settle on the bottom. As a result of the transformation of the detritus, the soil in the estuarine zone is enriched by organic matter. This soil generally consists of clay-silt material. The greatest enrichment takes place along sections of the river where the rate of flow approaches zero and in the region where the sea deepens.

What is taking place here is not a mechanical increase in the quantity of organic matter proportional to the supply of detritus, but rather a progressive increase in the mass of the accumulating organic substrate, giving rise to the vigorous development of organic life in the marginal marine plain. Where the sea deepens conditions are favorable not only for the accumulation of detritus, but also for burial of plankton remains.

Since all the large rivers have broad floodplains in their estuarine areas, where summer plankton proliferate, the water humus, according to Samoilov, must be of mixed type undergoing an increase in the plankton component.

The content of organic matter is highest in rivers of the equatorial zone and is also considerable in rivers of temperate latitudes, whose basins contain many lakes.

Microorganisms and enzymes cause gradual mineralization of unstable organic compounds in natural waters, while more stable groups become humified. In the water of the marginal marine plain, at the outer edge of the deltas, the amounts of organic matter and dissolved oxygen are lower than those in the estuarine channels. Here, the biogenic elements accumulate in the water, promoting local development of phyto- and zooplankton, while the more stable organic compounds, synthesized anew by microorganisms, accumulate in the sediments.

Even a brief study may serve to reveal the significant role of estuarine regions in the accumulation of the primary organic material for the formation of petroleum. In the vast expanse of estuarine regions conditions are favorable for the development of the biocoenosis, the accumulation and relatively rapid settling of muds enriched in organic materials, and the formation of various forms of sand bodies. All the necessary prerequisites for the subsequent generation, accumulation, and conservation of oil and gas exist here. The only exceptions are areas which have undergone repeated uplifting with resultant emergence and breaks in sedimentation.

Moreover, rivers build extensive alluvial coastal plains and deposit thick formations of terrigenous rocks in the zone transitional from dry land to sea. These sedimentary formations are often quite rich in dispersed and concentrated organic matter and also contain rocks with good reservoir qualities.

It is not by chance that the largest coal basins in the world are associated with alluvial-deltaic formations. Examples are the Donets, Kuznetsk,

Karaganda and other basins in the U. S. S. R., the Polish Silesian basin, the Com-
mentry basin in France, and the Appalachian, Kansas, Missouri and other
basins in the U. S. A.

In our previous works (1955—1970) we have established that the largest
oil deposits are regularly associated with deltaic and prodeltaic deposits
of paleorivers, the mouths of which underwent steady subsidence. This has
been confirmed both in the U. S. S. R. and in other countries, lending great importance to the detection of ancient estuarine zones within oil and gas basins.

*River Mouths*

The mouths of present-day rivers have been dealt with in numerous works
by Soviet and non-Soviet researchers. Samoilova's monograph (1952) provides a comprehensive review of the literature on this subject. Geologic-
geomorphologic descriptions of deltas are available in the works of V. V. Do-
kuchaev, N. Ya. Danilevskii, V. P. Baturin, D. V. Nalivkin, L. V. Pustovalov,
M. M. Zhukov, M. V. Klenova, V. P. Zenkovich and others.

A river, as it discharges into a water basin, forms a delta or an estuary.
The latter term refers to the single-channel mouth portions of a river devoid
of deltaic formations. According to a variety of characteristics, the following types of deltas are distinguished: infilling, cuspate, lobate, beaklike,
complex and bay (Figure 10).

Infilling deltas are formed when rivers flow into gulfs or bays. The fluviatile deposits fill them and then extend to the marginal marine plain. The
Don delta is an example: it was built up in a locality formerly occupied by
a gulf of the Sea of Azov, which had stretched far inland.

Cuspate deltas (for instance, those of the Volga, Lena and Ural) are typical
of open shores which are convex seaward.

Lobate deltas are formed when rivers deposit large solid loads and estuarine spits are formed in separate distributaries. These spits form peculiar lobes, protruding into the sea. The Mississippi delta is an example.

The beaklike deltas have the shape of a bird's beak. They usually consist of two estuarine spits. If the sea bottom is steep, such deltas may
sometimes develop into closed deltas.

Complex deltas are produced by a combination of two or more of the
above types.

Bay deltas are formed in cases when a river debouches into a bay or
lagoon separated from the sea by a barrier beach.

Intensive deposition of sedimentary material is quite characteristic of
deltaic regions; not only does this take place over vast water areas, but it
is also characterized by the rapid rate of growth of the shores and their
advance into the sea (Figure 11).

While studying the evolution of deltas, Samoilov proposed a genetic classification of river mouths. He divides them into morphogenetic types in the
order of their development and, consequently, in the order of relief complication (see Table 3).

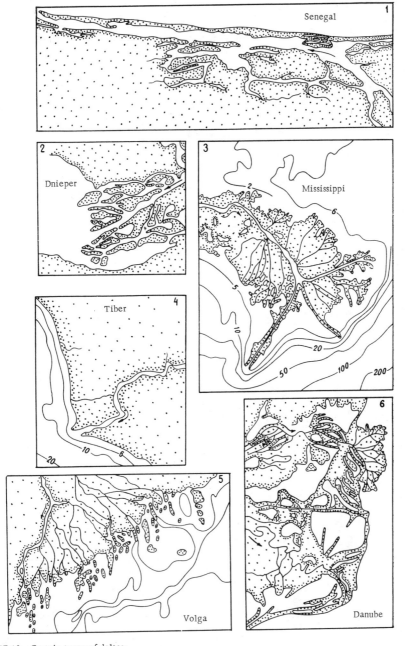

FIGURE 10.  Certain types of deltas:

1 – bay or closed;  2 – large channel (within a territory confined between the banks);  3 – lobate ("bird's foot");
4 – beaklike;  5 – cuspate (with numerous branching distributaries);  6 – infilling.

TABLE 3.  Morphogenetic types of mouths of present-day rivers (after Sarroilov)

| Estuarine section | Seaward part and margin of estuarine section | Estuarine marginal marine plain | Examples |
|---|---|---|---|
| **I. Single-channel:** | | | |
| a) with straight river bed | Large channel with wide estuarine spits (initial stage of mouth formation, breach) | Adjacent to deep water | Any new breach channel at the beginning of its existence. Vistula's Neufeuer mouth, Kargalinsk breach of the Terek (1914), Tiber (Fiumaro Grande), West Dvina |
| b) with straight bed, spit parallel to shore | Large channel with the wave-breaking spit parallel to the shore (closed mouth) | Adjacent to deep water | Kamchatka, Bolshaya (on Kamchatka), Senegal, certain channels of Niger, Rio Grande del Norte, Danzig Vistula (until 1845), Murray |
| c) funnel-shaped (tidal estuary) | Large channels bordering on permanent shoals or shoals that dry up during low tide | Adjacent to deep water, with primary subaqueous delta | Mezen, Khatanga, Elbe, Vezere, Thames, Mersey, Clyde, Seine, Loire, Gironde, Hooghly, Meghna, Congo, Saint Lawrence (with the relict Isle of Orleans), Hudson, Delaware, Potomac, Susquehanna, Magdalena, La Plata, Uruguay |
| **II. Double-channel:** | | | |
| a) with underwater delta | Two large channels separated by an underwater divide which is unstable and shaped by waves and currents | Adjacent to deep water, with developing subaqueous delta | Amur (bifurcation), Columbia, channels of many tidal deltas |
| b) with above-water (primary) delta | Two large channels separated by a divide which became a stable relief form – an island | Adjacent to deep water, with developing subaqueous delta and sloping sea bottom | Rion, West Dvina (13th century), Tiber (at time of formation of Sacra Island), Shatt al Arab, Yangtze |
| **III. Multichannel:** | | | |
| a) delta in the first stage of bay filling (within an area relatively confined between banks) | Large channels forming, with the aid of rills and spits, horseshoe shoals on the marginal marine plain: small lakes with islands (in the case of rivers with small solid load), or continuous shoals (in the case of rivers with large solid load) | Adjacent to deep water, with large rills and sloping sea bottom | Dnieper, Don, North Dvina, Pechora, Ob, Yenisei, Lena, Yana, Indigirka, Kolyma, Neva (?), Rhine, Mackenzie, Amazon |

| | | | |
|---|---|---|---|
| b) delta advancing into the sea c) lobate delta (sometimes secondary) | Relatively large channels with broad spits, usually becoming divided when the spits are breached; when a divide is formed on the leeward side the waves produce a spit from the alluvium not incorporated in the body of the delta | Adjacent to deep water and transitional to shoal, with unstable rills and sloping sea bottom | Lena, Yukon, Niger Kura, Ural, Dniester, Selenga, Kiliya delta of Danube, Po, Ebro, Hwang Ho, Mississippi |
| IV. Large insular: a) plain sloping seaward, with wide-meshed relief of the depressions between the raised levees of the distributaries: | Large, widely spaced channels which, if remaining in the same position for a long period of time, supply detritus for the formation of interchannel wave-breaking spits, sealing off limans | Adjacent to deep water | |
| 1) with lagoonal marine margin or lagoonal littoral area | | | Kuban, Nile, Danube (with deposits not yet filled in — "cells" of the relief) |
| 2) with predominance of uninundated areas | Large, widely spaced distributaries disappearing during frequent breaches of the river beds in the mouth area; they are divided by sand beaches with wave-breaking bars (marginal marine plain adjacent to deep water) or subaqueous and supraqueous vegetation (marginal marine plain adjacent to shallow water) | Transitional to shoaly | Amu Darya, Syr Darya, Terek, Colorado (North), Indus, Mekong, Irrawaddy |
| b) gently sloping plain with numerous distributaries branching off in the littoral part of the delta | In addition to several large distributaries there are very small channels with long narrow estuarine spits, which seal off shallow bays (kultuks) that develop into lakes and later, after shoaling, become flat islands surrounded by growing levees; bifurcation of the channels generally occurs through paired breaches of the bar placers | Shoals with rills marking the future network of seaboard channels | Volga, eastern part of Lena delta, deltas of Don and Dnieper in near future, Orinoco (?), Neman (?), Nogat delta of Vistula (?), Paraná |

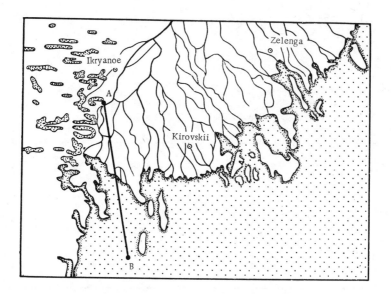

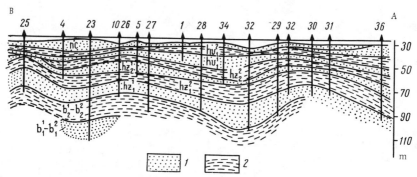

FIGURE 11.  Cross section of the edge of the Volga delta encroaching on the marginal marine plain (Rikhter et al., 1962):

1 — sand and coarse silt;  2 — clay and fine silt.  Age of deposits:  $b_1^1 - b_1^2$ — Early Baku;  $b_2^1 - b_2^2$ — Late Baku; $hz_1^1$ and $hz_1^2$ — Early Khazars;  $hz_2^1$ and $hz_2^2$ — Late Khazars;  $hv_1^1$ and $hv_1^2$ — Early Khvalynsk;  nc — Novo-Caspian and Recent alluvial deposits.

*Size and Structure of Deltas*

Samoilov describes the mouths of the world's largest rivers. Nalivkin's (1965) fundamental work on the facies clearly illustrates the huge size of some present-day deltas. The mouths of the Hwang Ho, Yangtze and certain other rivers form a gigantic band extending along the shores of the Yellow Sea for 1100 km. The width of the band of deltaic deposits, including the prodeltas, attains 500—600 km. The total area covered by the deltaic sediments is 500,000—600,000 km$^2$.

The delta regions of the Brahmaputra, Ganges and Mahanadi rivers extend along the shore of the Bay of Bengal for more than 50 km. The intradeltas occupy about 300 km, and the prodeltas — 100—150 km. The total area is about 150,000 km$^2$.

The intradelta of the Lena alone extends for 300 km, and is about 150 km wide. The configuration of the prodelta has not been established as yet. Nalivkin considers the possibility of a subaqueous link between deltas of the Lena, Yana, Olenek, Anabar and Khatanga. This territory extends along the shore for 900 km, and its width attains as much as 200 km.

The deltas of rivers in the northern latitudes — the Pechora, Ob, Yenisei, Kolyma and Mackenzie — are of approximately the same type. All of them have estuarine forms elongated in the direction of stream flow. Their intradeltas are 160—200 km long and 50—100 km wide. The area of their prodeltas is not known.

The well-studied Volga delta has more than 500 mouths, distributaries and channels. It is 250 km wide; with regard to the number of distributaries and the complexity of their ramification, it has no parallel. The mouths of the Volga and Ural are located in the Caspian Lowlands. Samoilov believes that the paleo-Volga and the paleo-Ural repeatedly formed a combined river system in their lower reaches and underwent considerable latitudinal displacement. This part of the Caspian Lowlands is rightly referred to as the deltaic alluvial plain.

On entering the Caspian Lowlands, the Volga sends out its first large distributary — the Akhtuba River — 450 km from the sea. Another major distributary — the Buzan — branches off 46 km north of Astrakhan. The entire Volga-Akhtuba floodplain abounds in oxbow-lakes, river channels and small creeks. The width of the Volga Valley ranges from 12 to 40 km, while the main watercourse is 0.6 to 2.2 km wide, its depth ranging from 2.5 m on sandbanks to 35 m on the reaches. In its lower reaches the Volga forms large and small distributaries. It flows into the sea through numerous mouths of distributaries and channels. The length of the major distributaries from their start to their point of debouchure into the sea is 70 to 110 km. Each distributary, as it approaches the sea, forms branches which in turn undergo further ramification. The upper part of the Volga delta contains 70 channels, the middle — 225, while in the lower coastal part their number exceeds 800. The distributaries and channels of the delta have different depths. The Volga's major watercourse has the greatest depths, up to 25 m. The depths in the distributaries are 6—10 m. They decrease to 1 m toward the sea.

The distributaries and channels continue into the marginal marine plain as shallow scours confined between lateral spits. Between adjacent spits,

there form gradually shoaling water bodies of shallow bays (or kultuks),
which serve as settling tanks for the inflowing river water.

Deltaic sediments are thought to account for about 25% of all sandstones.
Silt material (grain size from 0.05 to 0.1 mm) predominates (90—95%) in the
deposits of the Volga delta.  Coarser sand varieties are very uncommon.

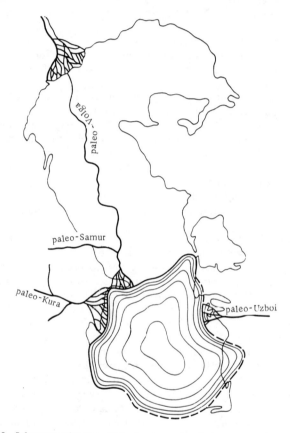

FIGURE 12.  Paleogeographic map of the Caspian region during the Pliocene (Baturin, 1937)

Baturin (1931—1947) has carried out an interesting study of the paleo-
Volga during the Neogene.  On the basis of a comprehensive petrographic
analysis, he established that the delta had formerly been located at the latitude of Apsheron (Figure 12); this study was primarily concerned with the
environmental conditions of deposition of the oil-producing formations of
the Apsheron Peninsula.  Baturin first investigated the sands of the producing formations.  He then determined the petrographic composition of
sands from all rivers of the Caucasus and Transcaucasus, as well as of the
Tertiary sandstones of these regions.  His comparison revealed that all the
sands of the Caucasus and Transcaucasus differ markedly from those in the
producing formations.

After Baturin had studied the sands of the present-day Volga delta and the Sea of Azov, it became evident that the deltaic sediments of the Volga, and especially those of its prodelta, are similar in composition to the arenaceous rocks of the Baku producing formation. The sands of the Mius liman on the north coast of the Sea of Azov were found to be of similar composition.

The example of the Volga serves to illustrate that a river, on entering the delta region, sends out distributaries meandering between the Baer knolls. These distributaries undergo further ramification into branches and shallow channels ("eriks") and form numerous exorheic lakes — ilmens. In the delta the solid load is distributed among the major channels approximately in proportion to their discharge. From the delta top to the dip of the sea bottom the content of the sediment fraction larger than 0.25 mm decreases, while that of 0.25—0.05 mm increases; the latter fraction accounts for 80—95% of the bottom sediments, so that the mean weighted diameter of the particles in the deltaic deposits is 0.15 mm. Banks of sand are formed along the channels and arms during floods. The structure and the development of the relief of the Volga delta are considered in great detail by Klenova (1948—1951), Samoilov (1952), Baidin (1956) and others.

Deltaic deposits are quite diverse and discontinuous, both in cross section and in plan. The courses and branches contain relatively well-sorted sand material. Sand mud accumulates closer to the banks where the rate of flow is slower. Constant change in the force and position of the streams in the delta is responsible for the great variety and displacement of the sediments.

Most of the sediments carried by rivers into the seas are classified as muds. They form the extensive subaqueous edges of deltaic regions and also provide a fill for the deltaic lakes (ilmens). The rate of sedimentation around the majority of deltas is rather high and changes abruptly as one moves away from the estuarine regions. The rate of formation of the deltas depends not only on the supply of sediments but also on fluctuations in the level of the basins into which the rivers flow. According to Klenova (1948), the Volga delta expanded 147—345 m per year during the period from 1817 to 1853; 47—720 m from 1853 to 1863; 272 m from 1864 to 1914; 530 m from 1914 to 1925; 119 m during 1934; 520 m during 1935; and 622 m during 1936. The fast growth during the last decades is due to the drop in the level of the Caspian Sea.

The lithologic composition and bedding of the deltaic deposits show great variation. Nalivkin (1956) considers the widely held opinion that deltas are characterized by coarse-grained sediments showing cross-bedding to be erroneous. A particular, relatively rare case may be mistaken for a regularity, thus leading to serious misconceptions. Nalivkin emphasizes the fact that a number of deltas have fine-grained sediments with regular bedding.

When studying the environments of estuarine sedimentation, one ought to bear in mind that the bulk of the sediments transported by rivers into bays and seas consists of mud, which forms extensive zones framing the prodeltas. It is in this mud that most of the primary organic material was concentrated, and when it sank to an optimal depth it developed into source beds in which oil and gas were generated.

Accumulation of organic matter of vegetable and planktonic origin in river mouths decreases from the outer edge of the delta toward the marginal marine plain. For instance, the average content of surface phytoplankton in the region opposite the Volga mouth drops from $100\,g/m^3$ near the delta edge to $0.5\,g/m^3$ near the shore of the Mangyshlak Peninsula (Figure 13).

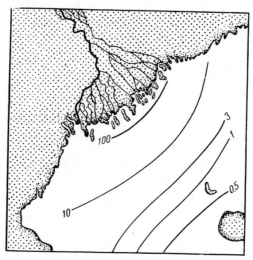

FIGURE 13. Distribution of the biomass of surface phytoplankton (in $g/m^3$) opposite the mouth of the Volga (Samoilov, 1952)

The marginal marine plain is characterized by the widespread occurrence of muddy ground which produces methane and hydrogen on decomposition of the organic matter. This process is arrested by the activity of bacteria which form a bacterial film. The latter serves as an excellent substrate for the proliferation of unicellular algae which synthesize new masses of organic matter.

The bedding of deltaic deposits is directly dependent on the site of their accumulation. In addition to fluviatile sediments, there are also lacustrine, palustrine, bar, and marine types of deposits, complicated by superimposed processes of flooding, shoreline migration and tectonic movements. The alternation of marine and terrestrial sediments is the most typical feature of deltaic formations.

Deltas of large rivers flowing in plains occupy enormous territories, and consequently, during flooding or temporary ingressions of the sea they become basins in which the entire area of the delta may be covered by arenaceous-argillaceous or argillaceous sediments. Repeated changes in such sedimentation conditions may lead to regular bedding.

Nalivkin (1956) noted a very important phenomenon which had not been discussed previously in the geological literature. This phenomenon occurs in the mouths of large rivers, especially on the surface of prodeltas in estuaries. Fluviatile sediments are often deposited within the prodelta and the adjacent regions and fresh- and brackish-water fauna develop. Such regions may extend for hundreds of kilometers. For instance, the Nile, with its relatively small delta, has a prodelta more than 100 km long and

400 km wide. During epochs of intense and protracted drought, the amount of fresh water entering the estuarine region drops radically, a normal marine salinity is restored and a marine fauna reappears.

The following is a description of a typical cross section produced solely by a change in climatic conditions. The base consists of a rather thick series of arenaceous-argillaceous deposits containing fresh- and brackish-water fauna; it is overlain by a thinner arenaceous-argillaceous formation with marine fauna. Such a cross section will, in all probability, be interpreted as resulting from a marine transgression. In point of fact, no transgression has taken place, and the sea level has not changed. Nalivkin terms this phenomenon "pseudotransgression."

The make-up of the prodelta fauna depends on the interrelationship between the river and sea waters. As the amount of fresh water decreases, a peculiar brackish-water fauna makes its appearance. In most cases it is mistaken for a lagoonal fauna, the implication being that the delta was replaced by a lagoon. This is not so, says Nalivkin. A delta is never replaced by a lagoon. In cases when fresh water is replaced by sea water, the fauna becomes of marine type, and the sedimentary deposits are then erroneously considered to be marine. To cite an example, the limestone and the clay intercalations containing marine fauna in the Donets, Moscow, Kizel and other coal basins are erroneously regarded as marine deposits formed as a result of delta subsidence and marine transgression. All these interlayers are normal deltaic deposits. They are the result either of a decrease in the supply of river water during dry periods or of intensified nearshore currents.

It is interesting to note that the French geologist Gignoux (1950) has expressed a similar opinion concerning the rhythmic structure of the coal-bearing deposits of the Franco-Belgian coal basin. He thinks that this structure is most probably due to periodic changes in climate, rather than to fluctuations in sea level caused by tectonic movements, as is commonly believed by those who adopt the classical approach to the formation of rhythmic sedimentary sequences.

The above statements have a special bearing on the reconstruction of ancient deltas. In Nalivkin's opinion, it is necessary to bear in mind, when studying the fossil fauna of Paleozoic deltas, that at that time certain marine genera of pelecypods, gastropods and even brachiopods and echinoderms started or attempted to adapt to life in both brackish and fresh waters. Therefore, their representatives, retaining their marine appearance, may be found in deltaic deposits.

We are paying such detailed attention to deltaic sedimentation because, as will be shown later, the major oil- and gas-bearing regions of the world are associated with ancient deltas. The deltas of humid zones are frequently characterized by yet another feature: widespread coal formations.

Very interesting in this respect is the Donets coal basin, which is one of the largest and long-existing Late Paleozoic paleodeltas. Its major coal-bearing series is of Middle Carboniferous age. It attains a thickness of 6—8 km and is composed mainly of arenaceous-argillaceous deposits with coal and limestone interlayers containing marine fauna. According to the structure and composition of the sedimentary deposits, it was a typical deltaic region which was fed by one or more rivers.

Nalivkin distinguishes three epochs in the evolution of this basin: 1) deposition of enormous amounts of fluviatile sediments (sand and clay); 2) formation of coal as the result of accumulation of plant remains; 3) formation of limestone.

During the first epoch the Donbas was the intradelta and partly the prodelta, with numerous river channels. During the period of coal formation, or more specifically, that of accumulation of plant remains, considerable areas of the delta were covered by tropical swamps. Finally, during the third epoch the basin was a shallow sea (10—30 m deep) into which no rivers debouched. In all probability, this is a case of pseudotransgression. However, subsidence of the delta below sea level does not lead to limestone formation but only to expansion of the prodelta. A necessary condition for the formation of limestone is the cessation, drying up or displacement of stream flow and not subsidence of the delta. Only in such circumstances will the water acquire normal marine salinity and will organisms forming limestones make their appearance.

Nalivkin therefore arrives at the conclusion that the Donbas was probably never subject to oscillations — uplift and downwarping — and only experienced alternations of pluvial and dry epochs.

We shall not attempt to assess the validity of this conclusion. However, one fact is certain: the Donets delta has repeatedly dried up at different periods of its evolution, leaving behind extensive swamps which existed for fairly long periods and served as centers of peat formation. Such periods were followed by inundation of the delta, accumulation of peat bogs, and subsequent periodic drying up and inundation. While this favored coal formation, the accumulation of bitumens was inhibited. We shall discuss this aspect in greater detail elsewhere in the book.

Yu. A. Zhemchuzhnikov, L. N. Botvinkina, P. P. Timofeev, A. P. Fcofilova, V. S. Yablokov, N. V. Logvinenko and others have ascertained that the various deposits of the Donbas coal-bearing formation belong to different facies. They include alluvial facies of both fluviatile and floodplain type, lacustrine, palustrine and deltaic deposits, sediments of semiclosed bays in the delta region or at a distance from it, bar deposits and, finally, various marine facies. All this shows how difficult it is to reconstruct the physiographic environment at a particular moment in the development of a sedimentary basin, even if, as in the case of the Donbas, there are good exposures and a dense network of boreholes. The paleogeographer will only succeed if he adopts a broad regional approach and accurately determines the genetic features of the deposits encountered.


*Ancient Deltas*


A comprehensive review of the distribution of the world's largest oil deposits reveals that most of them are associated with deltaic and prodeltaic sediments. Unfortunately, this conclusion, although of the greatest practical importance, is scarcely taken into account in prediction and the planning of prospecting operations. We shall try to furnish concrete examples to substantiate this conclusion.

Ver Wiebe (1959), a well-known American geologist, believes that the majority of oil deposits in the world are associated with ancient deltas. Rainwater (1964) states that traps located in large deltas of paleorivers which flowed into oceans, epicontinental seas or lakes are the most favorable ones for oil and gas accumulation. He points out that paleodeltas should be the primary object of search in petroleum exploration. On the basis of studies of the deltaic Booch sand in eastern Oklahoma, Busch (1965) maintains that the oil pools there are virtually unrelated to structural traps, and, therefore, exploration is based on the detection and tracing of the producing sands of the paleodelta.

Since 1955 the present author has published a number of works pointing out the high oil content in the ancient deltas of the East European Platform, West Siberia and other regions. Undoubtedly, deltaic deposits are quite common in the earth's crust, but it is difficult to detect them and very little is known about them to date.

Extensive paleodeltas dating back to the Devonian have been discovered in the states of New York and Pennsylvania. The maximum thickness of their sedimentary deposits exceeds 3000 m. These deltaic formations form a gigantic wedge. In the east they are generally colored red, suggesting a subaerial environment of deposition of sand and mud. Further west blue, gray, and green hues predominate, indicating either a marine or a brackish-water depositional environment.

Numerous deltas of Upper Carboniferous age have been discovered in the U.S.A. They are common in Pennsylvania, western Virginia and, especially, Alabama. In the southwest of Indiana and in Illinois Late Carboniferous (Pennsylvanian) sediments were deposited in swamps. The large number of coal seams separated by limestone intercalations bears witness to the frequently changing environments of sedimentation within the paleo-deltas. These environments must have been similar to those of the Donbas. Ver Wiebe points out that the vast alluvial plains of Pennsylvanian age in Illinois, Indiana, Missouri, Kansas, Oklahoma and Texas were similar to the present-day large alluvial plain of the Ob River in West Siberia. In Illinois, for example, the thickness of the ancient fluviatile sandstones is 1.5—3 m, but in the deepest portions of the river course it sometimes attains 30 m. The sandstone is overlain by indistinctly bedded shale of nonmarine origin. The shale is covered by limestone containing concretions and admixtures indicative of freshwater deposition. Higher up in the cross section one encounters clay and coal and then black shales representing the initial deposits of a transgressing sea. The shale is apparently the source bed from which oil later migrated into the adjacent porous reservoirs. These argillaceous rocks are overlain by compact chemogenic limestones.

There are thick series of Early Pennsylvanian deltaic deposits in southern Oklahoma and in northeastern Texas. Further north, in eastern Oklahoma and eastern Kansas, marine deposits alternate with terrestrial ones. The same pattern is observed in the northern part of Texas where there are extensive lowland areas. This is borne out by the numerous coal seams. Channel sands in the underlying deposits suggest the existence of an ancient river system. The shorelines of the ancient Cherokee Sea in southeastern Kansas are clearly delineated by long, narrow strips of oil deposits in the Greenwood and Butler counties.

The Garnett and Bush City fields are typical channel deposits. They are located in shoestring sand lenses lying near the top of the Cherokee shale formation, which are fluviatile sediments of Pennsylvanian age. Their length is about 25 km and their width locally reaches 800 m. The average thickness of the producing sandstone is 15—20 m.

In the same region there are oil pools associated with offshore bars which are most characteristic of the Bartlesville and Burbank sandstones of the Cherokee formation in the lower part of the Pennsylvanian. They are almost identical in structure to present-day bars of the Atlantic Plain. They also form strips whose length ranges from 5 to 8 to several tens of kilometers and whose width is 0.5—4 km (Figure 4). Their thickness varies from 10 to 45 m.

Among the known paleodeltas Ver Wiebe identifies one that was formed in southeast Oklahoma and southwest Arkansas during the Carboniferous. Throughout the Paleozoic, almost up to the end of the Mississippian, relatively undisturbed sedimentation occurred here. Then there was an abrupt change. Llanoria started to experience uplifting, and coarse clastic sediments accumulated, the beds dipping to the north and northwest of Louisiana. These rocks are termed the Pushmataha series. They were subdivided into six formations of which the lowermost is more than 1500 m thick. The section of the entire series features ten sandstone zones and the same number of zones of siltstone and shale. The overall thickness of the series is more than 4570 m and attains a maximum near the source area, decreasing to the northwest.

According to Ver Wiebe, the Pushmataha series was deposited in a delta environment. Many of the layers show ripple marks and contain fossilized remains indicative of shallow-water conditions. The depth of the water apparently did not exceed 60 m. There were numerous temporary marshes and lakes flooded by the sea. As one moves further north, the coarse clastic rocks typical of southeastern Oklahoma gradually give way to finer-grained varieties: siltstone and shale. In northern Oklahoma even limestone is present.

In the northern part of the Western Interior basin the oil and gas pools in Pennsylvanian rocks are generally controlled by a lithologic factor. Three principal types can be distinguished: lithologically sealed pools within lenticular sand bodies of offshore bars; lithologically sealed pools in channels of ancient rivers (the distribution of oil and gas in them is determined by the spatial disposition of the river bed); and lithologically sealed pools associated with the pinch-out of sandstones at the top of the uplifts.

The largest oil fields have been discovered in deltaic deposits of Cenozoic age. Included in this category, in addition to the above-mentioned delta of the paleo-Volga, is the ancient Mississippi delta.

### Ancient Mississippi Formations

The lowland alluvial plain of the Mexican depression extends north of the Gulf Coast for nearly 1000 km. Its area, about 80,000 km$^2$, was shaped by the Mississippi, Rio Grande, Colorado and other smaller rivers. In the lower reaches of the Mississippi, a vast delta was formed and extended into the sea in the wake of the receding shoreline.  The Mexican depression is made up of Paleozoic, Mesozoic and Cenozoic deposits represented mainly by terrigenous formations.  The overall thickness of the Mesozoic and Cenozoic deposits alone exceeds 10 km.  It is one of the largest regions in the world undergoing steady protracted subsidence.

According to American geologists, the origin of this depression dates back to the Permian.  This territory was presumably part of Llanoria during the Carboniferous.  Intensive downwarping of the depression can be traced to the Cretaceous and Paleogene periods.  This subsidence was accompanied by the formation of a series of step faults in the crystalline basement, along the scarps of which the depression subsided toward the Gulf of Mexico.

The famous Gulf Coast oil and gas province is located in the Mexican depression and includes the coastal plain and the continental shelf of the Gulf of Mexico.  It occupies the territories of Texas, Arkansas, Louisiana and Mississippi, and the greater part of Alabama, Georgia and Florida.  One third of the U.S. annual petroleum production is from this region.

The producing Jurassic, Cretaceous and Paleogene formations were deposited over the major part of the depression under alternating marine, coastal and lagoonal conditions against the background of the relatively rapid subsidence of the basin.  Salt-dome structures developed on the Gulf Coast and partly on the northern flank of the depression.

We will not list the characteristic features of oil and gas distribution in individual areas of this rich province, which some geologists regard as a global pole of oil and gas accumulation.  There are many publications on this subject, both in the Soviet and non-Soviet literature.  We shall only mention that, within the coastal stretch of the Gulf of Mexico and on the shelf, the oil deposits occur mainly in sand formations, constituting a broad zone roughly paralleling the present-day shoreline.  Various types of pools are encountered here: structural, lithologic, stratigraphic and lithostratigraphic. Lithologic pools are quite widespread, often including structural-lithologic and lithologically sealed types.  The famous East Texas oil field is a typical lithostratigraphic deposit sealed by an unconformity.

The largest known petroleum reserves in the U.S.A., with the exception of the oil deposits discovered in Alaska in recent years, are associated with the Tertiary deposits of the Mississippi delta.

According to Ver Wiebe's data, the origin of the Mississippi delta dates back to the Eocene.  In the south of Louisiana, at a depth of more than 4.5 km the Eocene formations consist of alternating deltaic and marine sediments. Sandstone lenses alternate with shales and thin beds of lignite.  The thickness of the Eocene sequence increases rapidly from north to south and attains a maximum of more than 1.5 km.

The Oligocene deposits also consist of alternating sands and clays of the deltaic type, with numerous limestone and marl intercalations.  The maximum thickness of the Oligocene in East Texas is about 360 m.  Miocene deposits account for most of the oil and gas production in southern Louisiana and southeastern Texas.  They are almost exclusively deltaic sediments. It is difficult to distinguish the Pliocene deposits from the Miocene ones. Consequently, they are regarded as a single Miocene-Pliocene series of deltaic sediments whose thickness is about 1220 m.  They account for a considerable part of the oil production in southern Louisiana and on the adjacent shelf.

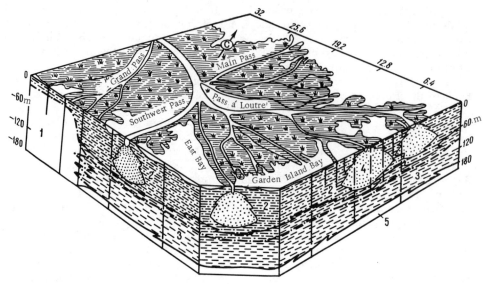

FIGURE 14.  The Mississippi delta (Fisk et al., 1955):

1 — deposits underlying deltaic formations; 2 — fluviatile clays; 3 — marine clays; 4 — sand lenses of the delta channels; 5 — marine sands.

The present-day delta of the Mississippi (Figure 14) is made up of Pleistocene and Quaternary sediments.    The river cut its course in the principal direction of flow of the meltwaters from the Quaternary glaciation. At present, the rate of subsidence of the delta is lowest along the north coast of the Gulf, where a lobate delta branches off, and increases toward the south. The rate of subsidence is 0.26 m per 100 years along the shores of southwestern Louisiana; 0.6 m for the headwaters of the distributaries, 1.8 m for the lower reaches of the South Pass (distributary); and 2.05 m for the lower reaches of the Southwest Pass.  The ancient deltaic alluvial deposits overlie older deposits of the same type, which may be dated as Early Cretaceous or even older.

The repeated shifts in the location of the mouth of the Mississippi gave rise to an alternation of deltaic and prodeltaic deposits forming the huge ancient deltaic alluvial plain rich in oil and gas.

### Occurrence of Oil and Gas in Sediments of Ancient Rivers

I. O. Brod, N. Yu. Uspenskaya and other Soviet geologists compare the regional structure of the Gulf of Mexico with that of the Caspian syneclise. K. K. Gostintsev and V. A. Grossgeim regard this comparison as being somewhat artificial, since these structures differ both genetically and morphologically.  We tend to support the first view and wish to point out the marked similarity in the evolution of the Mexican depression and the Caspian syneclise; the latter has not been covered sufficiently by deep drilling.

The common features of the two structures are their relatively stable and prolonged subsidence, the accumulation of huge masses of sediments supplied by rivers, and the presence of large commercial deposits of oil and gas. Once it has been established that areas rich in oil and gas are associated with the estuarine regions of the paleo-Mississippi and the paleo-Volga, there will be every reason for believing that no less significant oil- and gas-bearing regions occur in the paleodeltas of the Ob, Angara, Irtysh and other rivers which have also changed their location. Later we shall discuss the paleorivers of West Siberia. In passing, we shall merely recall the comment made by Nalivkin that quite a few deltaic deposits have not yet been recognized as such, and have been included in the general list of lagoonal and terrestrial deposits. The situation is even less clear in the case of pro-delta formations.

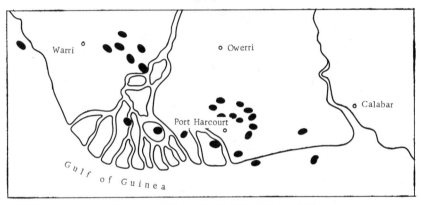

FIGURE 15.   Location of oil and gas fields in Nigeria (Bakirov et al., 1971)

There are numerous other examples of oil- and gas-bearing regions associated with ancient deltas. Rich deposits occur in the Orinoco River delta in Venezuela, and in the deltas of the Chindwin and Irrawaddy rivers in Burma, which are of the same age and are known for their large petroleum reserves. In recent years several oil fields were discovered on the coast of Nigeria and in the Gulf of Guinea (Figure 15), at the mouths of the Niger and Benue rivers. The first deposit was discovered in 1959, and in the following decade 70 fields began to be exploited, the annual production amounting to 71 million tons by 1970. The producing horizons are sand deposits of Miocene, Paleogene and Upper Cretaceous age. The initial yield of the first wells ranged from 100 to 300—400 tons per day. In the Gulf of Guinea, some 10—20 km offshore, several marine oil and gas areas were discovered; they were found to be more productive than those on dry land.

Central Oklahoma may serve as an example of the widespread occurrence of oil pools in sand formations of ancient deltas of Mississippian age. Here, in the Booch sand, there are numerous pools located in distributary and channel deposits of the paleodelta of a large river which debouched into a sea formerly occupying the McAlester basin (Figure 16).  The area of this paleodelta exceeds 5200 km$^2$.  The thickness of the Booch channel sand ranges from 6 m to more than 80 m.  Mainly clay deposits accumulated between the river courses and channels.

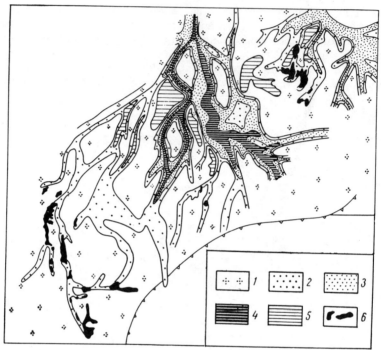

FIGURE 16.  Distributaries of paleodelta in Lower Carboniferous time in central Oklahoma (Busch, 1959):

sand thickness (m):  1 — from 0 to 6;  2 — from 6 to 20;  3 — from 20 to 36;  4 — from 36 to 72 or more; 5 — clay;  6 — oil pools.

As is quite evident from the map showing part of the ancient delta, not all of the channel sandstones are oil-saturated; only those located in the prodelta zone fall in this category.  This is probably due to the fact that the oil migrated from sediments in the deeper portions of the marine basin toward the coast and accumulated in the very first traps it encountered on its way.  Thus, it is very important to have knowledge of the position of the deltaic sand formations relative to that of the source beds.

Figure 17 shows a portion of the above-mentioned delta in central Oklahoma, the distributaries and channels of which contain oil.  In this particular case the location of the oil pools is independent of the structural factor.

The mouth of an Early Cretaceous paleoriver in Alberta (Canada) is another example of oil accumulation in the alluvial deltaic sands.  Here the fluviatile sands fill a broad erosion valley incised in Paleozoic carbonate rocks (Figure 18).

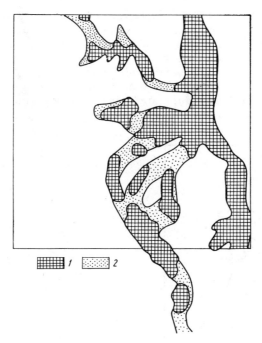

FIGURE 17. Oil pools in one of the channels of the ancient Booch River, Oklahoma (Busch, 1959): sandstones: 1 — oil-saturated; 2 — nonproducing.

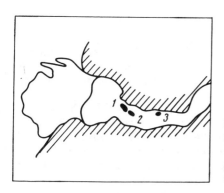

FIGURE 18. Early Cretaceous river mouth with producing fluviatile sand, Alberta (Conybeare, 1964):

Oil fields: 1 — Bellshill Lake; 2 — Thompson Lake; 3 — Hughgenden.

Despite all the difficulties involved in detecting fluviatile sandstones, they should not be ignored, since they often serve as good reservoirs for oil and gas. The search for oil in such sandstones may be rewarding, provided one is familiar with their distribution pattern. When an oil pool is discovered in them, the main problem is to determine the most appropriate location of the next well. This calls for information on the trend of the sandstone, as well as on changes in its porosity and permeability. In the absence of accurate knowledge of the genetic features of the sandstone, the

shape of the sand bodies and their relationship to possible source beds, it
is difficult to establish the optimal location of either test or producing wells.

Most alluvial deposits are either linear or meandering. Prodelta sediments have a rather irregular distribution. The lines of strike merely indicate the general outlines of the area which may contain oil and gas deposits (Figure 19). In order to be able to make more accurate recommendations, one requires knowledge of the structural, hydrodynamic and other conditions attending the formation of the sandstone.

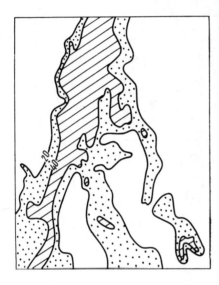

FIGURE 19.   Isopach map of alluvial oil-bearing
Frio sands of the Seeligson field, Texas (Conybeare, 1904)

Many oil deposits in Indonesia, Peru, India, Pakistan and other countries
are associated with deltas of Cenozoic rivers. Thick producing deltaic formations of Cretaceous age were discovered in the U. S. A. and in western
Canada.

Apparently, it is not by chance that most of the coal-bearing formations
in Europe and Asia, as well as in North America, are located in deltaic regions of sedimentation. Not infrequently these strata develop into oil and
gas fields along the strike. A possible example is the Angara-Chulym depression in the West Siberian Lowlands, which is rated among the world's
largest oil and gas basins.

The coal-bearing alluvial deltaic Jurassic formations of this depression
develop into oil- and gas-bearing nearshore marine formations toward the
more downwarped central areas of West Siberia. While studying the lithofacies of the coal-bearing deposits, Timofeev (1970) discovered that the ancient land area from which the clastic material was carried into the Angara-
Chulym depression had occupied three adjacent regions: the Baikal Range
in the southeast, the Northern Baikal Range and the Lower Tunguska-Stony
Tunguska interfluve in the northeast, and the Kuznetsk Alatau together with
the spurs of the Western and Eastern Sayan in the southwest. To the west
and northwest of the depression there was an extensive alluvial deltaic region
where the ancient rivers discharged their load.

According to Timofeev, the paleo-Angara River was once the largest and principal water way.  North of Krasnoyarsk it flowed from east to west, being fed by large and small tributaries.  The paleo-Tunguska was the second largest river and evidently had two sources.  The southern source incised its course in granitic massifs and schist, while the northern source cut through trapps and sedimentary rocks of the Stony Tunguska-Lower Tunguska interfluve.  The paleo-Chulym was the third largest river; it had its source in the Kuznetsk Alatau where it passed through granitic and volcanic massifs, as well as sand-silt formations.  All these rivers draining into the marine basin that then occupied the greater part of the West Siberian plate terminated in vast intradeltas or prodeltas.

The subaerial portions of the paleodeltas with their numerous distributaries passed into their subaqueous continuation in the shallow marginal marine plain where the depths remained insignificant over a distance of hundreds of kilometers from the shore.  This is indicated, in particular, by the vast area of alluvial deltaic, lagoonal, nearshore and shallow marine deposits.  In the estuarine regions, alongside various accretion forms of sand formations, even greater amounts of argillaceous sediments were deposited, which could well become source beds or serve as cap rocks.

The terrigenous material carried by the rivers was distributed in accordance with the dynamics of the aqueous environment.  Beyond the highly dissected coastal strip, sand deposits of the river beds were carried over distances of tens or hundreds of kilometers, sometimes acquiring the shape of channellike bodies and sometimes that of the bodies extending in the direction of bottom currents.  It is not surprising that in the  Shirotnaya Ob region the reservoir rocks represented by sand-silt varieties show typical wave-cross, criss-cross or horizontal bedding.  The sand members, which often are lense-shaped, alternate with layers and intercalations of mudstones of floodplain lagoonal-bay and nearshore marine facies.  In places there occur bodies that seem to be beach sands with indistinct bedding.

All features typical of estuarine zones of ancient, fairly large rivers are noted in the southeastern part of West Siberia, where the following important oil fields were discovered: Samotlor, Mamontovo, Ust Balyk, Pravda, etc. Further detailed studies and the tracing of the extent of deltas of other ancient rivers (aside from the paleo-Angara, paleo-Tunguska and paleo-Chulym) which debouched into the Mesozoic sea, will undoubtedly lead to the discovery of several large oil and gas deposits.

If more attention were paid to the genetic implications of producing formations during studies of individual oil pools and fields, this might furnish much additional information on the location of many of them in the delta zones of ancient rivers.  This is borne out by the experience gained in studying the producing formations of the Visean stage (Lower Carboniferous) on the East European Platform, which will be discussed later in this book.

## 5.   REEF FORMATIONS

When considering the oil and gas prospects of the coastal zone, one should not ignore reef formations which develop as a result of the life activity of colonial organisms.  Ancient reef formations, like sandstones, are excellent natural reservoirs for the accumulation of oil and gas.  A large number of hydrocarbon deposits in the U. S. A., Canada, Mexico, Libya, Iran, Syria and other countries are associated with reefs.

In the Soviet Union the first oil pool in a reef body of Permian age was discovered in 1929 in the vicinity of Chusovskie Gorodki in the Western Urals.  This was followed by the discovery, in 1932, of Bashkiria's renowned Ishimbai field consisting of several interconnected reef bodies constituting a single deposit.

To date more than 20 oil and gas pools occurring in the zone of reef development have been discovered in the Belski depression of the Ural fore-deep alone.  These barrier reefs extend along the ancient shoreline of the Early Permian marine basin.  They generally have the shape of conical elevations and overlie the basal deposits of the Upper Carboniferous.

Oil-bearing formations were detected in the Kama-Kinel depression system, in the Timan-Pechora basin and in other regions of the U. S. S. R.  Interest in the conditions of formation and the distribution of ancient reefs has increased in recent years.  Nevertheless, it must be admitted that in the Soviet Union they have not yet been studied sufficiently from the paleogeographic aspect.

It is commonly assumed that reefs form in the shallow, warm, clear waters of marine basins.  Reef-building organisms, such as corals, algae, bryozoans, crinoids, echinoderms and others, show optimal development in shallow seas with an average annual temperature of 20°C.  Most reefs are formed at depths of 20—40 m, mainly along the shores receiving no river runoff.  Exceptions are the reefs which originate in submarine volcanic craters once they have been raised to a suitable water depth.

Nalivkin (1956) enumerates the following characteristic features of ancient reef bodies: predominance of massive, fairly pure organogenic limestones, with bedded limestones along the margins of the bodies or within them; irregular conical, knolly or convex lenticular shape; distinct, often sharp outlines; regular distribution patterns in most cases associated with tectonic features.

We believe that the distribution patterns of reefs are controlled by paleogeographic no less than by structural conditions.  The reef formations themselves are reliable indicators of a specific physiographic environment of sedimentogenesis accompanied by biogenesis.

Reefs are divided into four main types: shore, barrier, atoll and elevated. Some investigators distinguish between platform reefs and geosynclinal reefs, the latter including the reefs of anticlines and those of volcanoes.

Shore reefs form on or near the shore.  They usually occur as narrow discontinuous bands whose width ranges from several tens of meters to several kilometers.  The limestone thickness in these reefs does not exceed several dozens of meters.

Barrier reefs occur along the inner margins of the shelf and are located at distances of up to tens of kilometers from the shore.  They form a nearly continuous band, up to several kilometers in width, along the shoreline.  Barrier reefs are not usually thick, but in the case of prolonged shelf subsidence their thickness may attain hundreds of meters.

Atolls are ring-shaped coral islands resulting from the subsidence of volcanic islands. They are not very large in area, but may be up to hundreds of meters thick.

Elevated reefs are barrier or atoll reefs raised above sea level. Their height sometimes reaches a few hundred meters.

Although the above classification relates to present-day reefs, the distribution of buried reef bodies corresponds to similar or analogous physiographic conditions. Therefore, one can predict the distribution of ancient reef formations primarily on the basis of paleogeographic evidence.

Some ancient reef bodies presently appear as relict knolls made up of organogenic limestone and dolomite. Such knolls are widespread on the water-divide plateaus in Bashkiria where they are known as "shikhan." They are conical elevations of Early Permian limestone rising above the surrounding relief. The limestone is not bedded, but sometimes dips parallel to the slopes of the knolls.

The best studied reef formations in the Soviet Union are those of the Upper Paleozoic, which can be traced along the western slope of the Ural Mountains. Their structure and composition have been dealt with in numerous specialized works, but their physiographic features have not been covered adequately.

The Tampico region in Mexico is a typical example of an oil deposit occurring in a reef formation (Figure 20). Here the oil pools associated with reefs of Cretaceous age extend as a chain along the present-day northwestern shore of the Gulf of Mexico, indicating the general outlines of the ancient sea which filled the Mexican foredeep during the Cretaceous. The producing horizons are fractured and, in places, cavernous limestones of the El Abra formation.

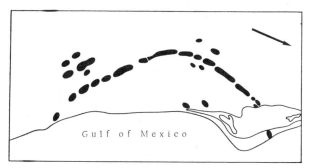

FIGURE 20. Location of oil occurrences associated with reef formations in the Tampico region (Guzman and Mina, 1952)

Reef limestones and dolomites usually differ from most other carbonate rocks in that they constitute solid, massive bodies even at the sedimentation stage. The reef formation itself often displays facies variations from the core toward the flanks. The reef core consists of secondary amorphous dolomite. Here one encounters many cavities and fractures. On the reef

slopes there are numerous fragments of fossilized remains of organisms
which accumulated around the core.  Reefs also incorporate detrital material brought by the waves and deposited in cavities and channels.

Although organic remains usually make up the bulk of the reef, in some
cases detrital material accounts for as much as 90% of the total mass of
the bioherm.  According to Sharma (1968), a considerable part of the detrital
material results from the disintegration of the primary reef formation.
The most characteristic feature of reef rocks is the almost total absence of
extraneous material.  Consequently, they are pure carbonate rocks and contain a negligible amount of insoluble residue.

Murray (1968) points out that the oldest Paleozoic stromatoporoid reefs
are of Middle Ordovician age.  By the beginning of the Silurian they were
already widely distributed.  The stromatoporoids continued to be active
reef-building organisms in the Devonian.  In places solenoporaceae and
tabulate corals and, to a lesser extent, oncolites, brachiopods and echinoderms acquire significance as "rock-formers."

Algal-bryozoan reefs (oncoids) differ from coral reefs in that their formation does not depend on the water temperature; they occur in both tropical and polar seas.  They can form over a wider depth range than coral
reefs, from several meters in the surf zone to the depth of penetration of
light (100—150 m).  According to Nalivkin's findings, the salinity of the water
has a considerably lesser effect on their formation.  While coral reefs are
associated with normal salinity, algal-bryozoan reefs develop fairly well
also in endorheic brackish-water basins and even in fresh-water and bitter
lakes.

Studies of present-day reefs have shown that their ecological features
are essentially determined by the wind rose and the consequent predominant
wave direction.  In the opinion of Ingels (1968), it is the waves that are responsible for the distribution pattern of the organisms and sediments and
the direction of principal growth of the reef body, which, in turn, determines
its shape.

Consequently, the shape of the reef depends on the force and direction of
winds, wave motions and currents and also on the size of the space available
for their development.  Murray has shown that the currents, reaching the
shallow bottoms of shelf seas, start to act on the reef from the moment of
its origin and thus determine its subsequent orientation in plan, while wave
action is responsible for its shape during the later stages of reef development.  In narrow elongated shelf seas the direction of currents is often at
variance with the predominant wind direction; in such cases a wide variety
of reef shapes may arise.  Unfortunately, very little attention is paid to this
aspect in studies of buried reefs.

The size and shape of reef formations vary widely, their thicknesses
ranging from a few tens of meters to 2 or 3 kilometers.  The Great Barrier
Reef (Australia) has attained gigantic size.  It is about 2000 km long, 200 km
wide and at least 400 m thick.  Numerous reefs of the Indonesian Archipelago, the Pacific Ocean and Florida extend for hundreds of kilometers.  Small,
thin, lenticular reef formations are also known.

The clue to the formation of reefs in the tidal zone should be sought, according to Nalivkin, in the composition of the rock-forming organisms.

Massive reef-building corals and calcareous algae are indicative of formation in the tidal zone, while brittle bryozoans and ampullaceous and lichen-
like algae build reefs at depths of several tens of meters.   Another characteristic, revealed by drilling data, is the minor role of coral limestone in
reef formation; this holds both for modern and, especially, ancient reef limestones.

Ancient reefs are usually highly altered and recrystallized; dolomitization is very common.   In addition to the reef proper, a reef complex also
includes carbonate deposits of the forereef and backreef.   The forereef dips
steeply away from the reef and forms a lenticular or wedge-shaped deposit.
The steep dip is due principally to the deposition of talus on the reef slopes,
rather than to tectonic causes.   Backreef sediments often show regular
bedding and a fine-grained texture; they generally consist of calcareous
mud, algal limestone and dolomite.

During the process of their growth, the reefs rise considerably above the
surrounding portions of the sea bottom.   As noted by Sharma, this compli-
cates the stratigraphic study of ancient reef bodies.   Lagoons could form
either inside the reef or on one of its sides.   In any event this produced
three different but coexistent environments of sedimentation and development of organisms.

Such a distribution of facies environments could be conducive to the accumulation of the primary organic matter from which hydrocarbons are subsequently formed.   The calm waters of lagoons, beyond the sphere of hydrodynamic wave action, usually provided the necessary reducing medium.   Nalivkin (1956) gives a detailed review of the formation and distribution of
coral islands.   The following are their principal features: 1) location in
tropical regions;  2) occurrence as long, narrow strips;  3) dependence on
the character of the shore.

The mentioned work presents a theory of reef development, which can be
summarized as follows:

1) The thickness of the limestone formations, both reef and associated
nonreef, reaches 2—3 thousand meters.   Such thickness can only be the result of prolonged subsidence;

2) Elevated reefs result from repeated uplifting of the base;

3) Some modern reefs grow under conditions of an immobile base; however, this immobility is only temporary;

4) The formation of the ice sheet led to a slight drop in the ocean level,
which is reflected in the growth of the reefs;

5) In a number of regions, submarine folds and tectonic blocks, which alternately rose and subsided, played a major role in the formation and growth
of the reefs;

6) Many reefs originate on submarine volcanic cones that reach the sea
surface.

The location of the majority of present-day reefs coincides with that of
Quaternary ones.   Pliocene reefs may be found slightly further north than
modern or Quaternary ones.   They are encountered on the south coast of
the Mediterranean.   The distribution limit shifted far to the north during
the Miocene.   Coral reefs are developed along the North African coast, in
Italy and Israel, along the north coast of the Mediterranean and in Asia Minor.

During the Cretaceous the reefs advanced even further north and in the Jurassic reef limestones developed on a colossal scale. They have been found in the Pamirs, in Tadzhikistan, on Kopet Dagh, on the northern slope of the Caucasus, in Crimea, north of the Alps, in England, etc. The Tyrol dolomites of Triassic age are reef limestones.

The Paleozoic reefs in the Soviet Union and North America have been fairly well studied in view of their oil and gas prospects. Best studied in the U. S. S. R. are the Upper Paleozoic reefs of the Ural zone. They can be traced south of the Northern Urals (in the valley of the Pechora River) along the entire western slope of the Ural Mountains (Figure 21).

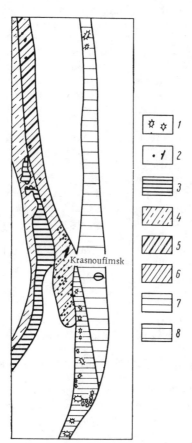

FIGURE 21. Distribution of Upper Paleozoic reefs in the Ural region (Nalivkin, 1956):

1 — Duva reefs of Sakmarian age; 2 — Sarga reefs of Artinskian age. Regions of occurrence: 3 — Sylva reefs of Late Artinskian age; 4 — probable development of Sylva reefs; 5 — Sarga reefs; 6 — buried Sarga reefs; 7 — Duva reefs; 8 — probable development of Duva reefs.

The overall extent of the reef regions is estimated at 1000 km; the width varies from 3 to 18 km; some of the reef formations attain a thickness of 1200 m. The vast majority of the reefs are situated at the boundary between bedded limestones and argillaceous-arenaceous deposits extending eastward, i. e., toward the shoreline. In contrast to present-day reefs, the Upper

Paleozoic barrier reefs of the Urals were separated from the continent by
considerably larger lagoons.   This circumstance apparently played a significant  role  in the accumulation of oil and gas within these reef bodies.

The structure of the producing reef formations of Ishimbai and other
regions has been studied in great detail and described by A. A. Trofimuk,
A. N. Dubrovin, N. P. Gerasimov, D. V. Nalivkin, V. P. Maslov, G. I. Teodorovich, and others.

In the U. S. A., oil- and gas-bearing Ordovician and Silurian reefs are
known to occur in Kansas and Illinois, in the neighboring states, in the Michigan basin and in other regions.   Oil deposits were discovered in Devonian
reef formations in the Judy Creek region, Alberta, Canada.

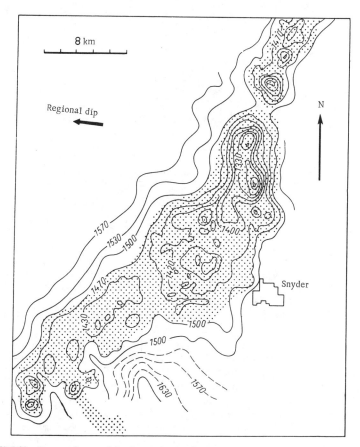

FIGURE 22.  Structural map of the Scurry-Snyder field.  Structure; contours show configuration
of top of reef limestone (Levorsen, 1970)

Among the largest reefs of Pennsylvanian age in the U. S. A. is the Scurry-
Snyder reef (Figure 22) in western Texas.   It overlies Mississippian limestones which are, in turn, underlain by Ordovician carbonate rocks.   The

hydrodynamic link between the reef series and the underlying limestones is regarded as the cause of petroleum accumulation in the mentioned reef, whose reserves are estimated at 145 million tons. Other Pennsylvanian reefs of this region, separated from the Ordovician sedimentary deposits by impermeable argillaceous rocks, are nonproducing. Figure 23 shows a cross section of a typical oil pool associated with the reef in the Scurry region.

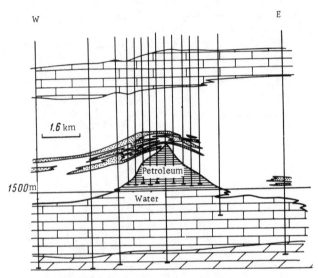

FIGURE 23. Section through the North Snyder pool in western Texas (Levorsen, 1970). The trap is an organic reef formation overlying Pennsylvanian limestone.

In the U. S. S. R., in addition to the Permian reef formations, there are also productive Devonian barrier reefs in the Kama-Kinel depression and in the Timan-Pechora basin. The Khilovskii oil field, for instance, is an erosion remnant of a reef body of Famennian age, which formed on the flank of the Kama-Kinel trough system. Starting from the Tournaisian an envelope structure was formed over it. The size of the structure as measured along the top of the Pashiya horizon is 6.3 by 1.5 km. The angle of dip of the northern flank is 11°30', and that of the southern flank — 1°40'. The amplitude of the southern flank is 32 m, while that of the northern flank is 134 m.

Structural indications exclusively led to the discovery of 23 oil and gas fields within the Kinel-Samarka system of linear dislocations, within which the Khilkovskii reef is located. The reef-type Khilkovskii uplift is, in all probability, not an isolated feature; a detailed reconstruction of the paleogeographic environments during various phases of the evolution of the Kama-Kinel trough system might help to reveal other similar structures.

The investigations of A. I. Antropov, M. M. Grachevskii, M. F. Mirchink, O. M. Mkrtchyan and others have already established the presence of various reef formations in the Late Devonian deposits of the Kama-Kinel trough system.  Unusual carbonate deposits of barrier reefs were detected on the flanks of the trough system (Voyady-Arlan-Dyurtyuli, Shumovskii, etc.).

The reef formations, in turn, are indicative of the paleotectonic and paleo-geographic setting at the time of their formation.  The shoreline of the Late Devonian marine basin of the Volga-Ural region was very complex.  In this region the conditions of sedimentation and the habitats of the organisms were markedly diverse.  Carbonate sediments with abundant reef-building fauna and calcareous algae were formed on the uplifted bottom areas.  Argillaceous (often bituminous), finely bedded limestones alternating with dark-gray mudstones and marls were deposited in the depressions.

Above we have cited some examples of giant oil fields outside the U.S.S.R., which are associated with ancient reefs.  We should now ask ourselves why such unique oil reserves occur within reef bodies.

Sorokin (1971) pointed out that researchers have long been aware of the abundance of life in zones of modern coral reefs, which are surrounded by poorly productive tropical waters containing a minimal plankton biomass and virtually devoid of salt-water biogens.  Large masses of water poor in organisms flow above the reefs as a result of tidal currents.  One might assume that the reefs rich in biocoenoses are continually losing their organisms and biogens.  However, this is not the case.  Not only are the coral reef biocoenoses preserved, but they may even be regarded as the most productive biocoenoses which have ever existed on earth.

Sorokin, who studied the production of modern coral biocoenoses, was greatly impressed by the abundance and activity of the bacterial population in the water layer and in the bottom sediments.  His observations have revealed the tremendous role played by the microflora in the production processes occurring in the marine waters adjacent to the reefs.  Both in clean and contaminated reef atolls the production and biomass of the bacteria may exceed those typical for the waters and sediments of the open ocean by a factor of several hundreds.  The daily production of the phytobenthos and bacteria in the sediments here reaches several percent of their total organic content, which points to exceptionally high rates of the biological cycle.

Could this not be one of the explanations for the enormous petroleum accumulations which are found in some reef bodies? It appears that an essential prerequisite for the formation of huge accumulations of oil in ancient reefs, as in the case of the richest oil fields associated with sediments of deltas and prodeltas, is an exceptionally high content of organic matter in the sediments of the relevant regions, which serve as source beds.

It is very difficult to detect buried reefs; their presence is usually revealed by sinking wells in local structures.  The search for reef traps differs little from exploration for local uplifts.  Seismic surveys are the most effective tool in poorly studied regions.  If the carbonate complex contains rocks differing in density, the seismic data may be helpful in determining the location and general shape of the reef formation.  Finer morphological features of the reef, its lithologic composition, physical properties and relationships with the enclosing rocks, as well as the content of oil and gas, can be revealed after the drilling data are processed.  Kuznetsov (1971) offers certain recommendations on the use of geophysical data for the exploration and study of reef formations.  However, the technique employed in the search for buried reefs as yet lacks a scientific basis.

Paleogeographic studies may make an essential contribution toward developing such a technique. Since the general distribution patterns of reefs are known, it is of prime importance to trace ancient shorelines and the character and regime of the marine basins themselves, i. e., to delineate the most likely zones of reef formations.

## 6.  OCCURRENCE OF OIL AND GAS IN THE COASTAL ZONE

It is an established fact that the coastal zone is the region characterized by the most vigorous interplay between the lithosphere, biosphere and hydrosphere, the greatest mechanical and partly chemical differentiation and accumulation of sedimentary material, and the highest content of organic matter in the sediments. Considerable reserves of fossil fuels occur in this zone. The latter circumstance is most important and is corroborated by the examples given below.

As already mentioned, oil and gas pools are particularly widely distributed in coastal areas which were affected by deltaic and prodelta sedimentogenesis. We shall first refer to some examples outside the Soviet Union.

The famous Gulf Coast oil and gas basin in the U. S. A. is considered by Veber (1956) as an example of a basin where oil pools occur in the zone of transition from marine facies to continental facies. He superimposed on the facies-geological cross section the location of oil pools within individual stratigraphic horizons of the Cenozoic (Figure 24).

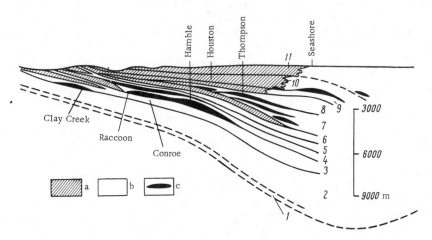

FIGURE 24.  Cross section of the Gulf Coast depression (after Meyer with additions by Veber, 1966):

Facies: a — continental; b — marine; c — oil pools (not shown in the Wilcox group). Eocene groups: 1 — Midway; 2 — Wilcox; 3 — Lower Claiborne; 4 — Upper Claiborne; 5 — Jackson; Oligocene formations: 6 — Vicksburg; 7 — Frio; Miocene; 8 — Discorbis zone; 9 — Catahoula; 10 — Pliocene-Miocene; 11 — Pliocene, Pleistocene, and Recent sediments.

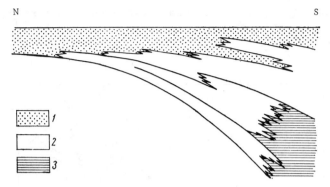

FIGURE 25. Section showing facies changes in the Upper Oligocene formations of the Gulf Coast (Lowman, 1949):

Facies: 1 — continental; 2 — neritic; 3 — bathyal

Figure 25 presents a section showing facies changes in the Upper Oligocene formations of the Gulf Coast in the direction of the regional slope, i. e., toward the modern gulf.  It depicts the transition from continental facies to marine facies (first neritic, then bathyal).  Individual advances and retreats of the facies boundaries can be observed against the general background of this transition.  On the whole, however, the marine facies gradually recede southward in the direction of the Gulf of Mexico.  The oil-bearing zones shift in the same direction (Figure 25).  Veber attributes this shift of the oil-bearing zones to the change in location of the facies environments favorable for oil accumulation.  In the process of this migration the oil- and gas-bearing zones are always distributed within the same depth intervals: from 2000 to 4000 m in various stratigraphic complexes.  Veber poses the question whether these depths could be optimal for the formation and accumulation of oil.  We believe that this question has been answered by Vassoevich (1967) who discusses the major phase of generation of oil and gas.

According to Veber and a number of other American geologists to whom he refers, facies environments favoring the generation of oil exist in the neritic region, mainly within the coastal zone adjacent to the region of continental facies.  The latter in this case are deltaic formations associated with the paleodeltas of the Mississippi, Colorado-Brazos, and Rio Grande rivers which have debouched into the Gulf of Mexico since the Paleogene or, possibly, even since the Cretaceous.

Murray (1960) indicates that the shift of the oil-bearing zones toward the Gulf is accompanied by a shift of the axis of maximum sedimentation, i. e., the regions of maximum subsidence of depressions also migrate.  According to all available data, the present-day conditions of sedimentation in the coastal zone of the Gulf of Mexico differ little from the depositional environment of the sediments of the Cenozoic producing formations.  The strike of the Cenozoic zones of oil and gas accumulation coincides with the present-day shoreline of the Gulf (Figure 26).

The entire shoreline of the Gulf, extending for 1000 km, is made up of an almost uninterrupted chain of lagoons and lagoonal bays.  The Madre lagoon,

200 km long and 4—8 km wide, has an average depth of approximately 0.8 m.
The bottom sediments here consist mainly of fine-grained sand, siltstone,
and clay.  This lagoon is one of the most productive in the world in terms
of the abundance of its flora and fauna.

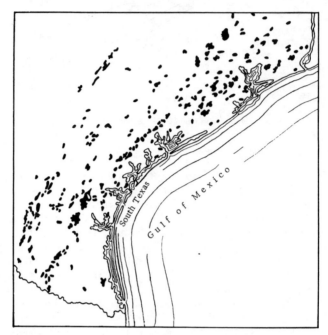

FIGURE 26.  Location of oil fields on the Gulf Coast

Veber also mentions other semi-landlocked and landlocked embayments
on the Gulf Coast.  He believes that the lagoonal facies is more favorable
than the shelf facies for hydrocarbon accumulation.  In regard to the accumulation and transformation of the original organic material, the inner portion of the barrier beach zone may be considered together with the lagoonal
zone as a single facies complex, both being equally favorable for the formation and accumulation of oil.  Veber is of the opinion that the sediments of
the adjacent shelf portion, being potentially productive, belong to this complex.
    He emphasizes that oil formation may take place, provided that the sedi-
ments of the lagoon — barrier beach — adjacent portion are constantly under
water and are not subjected to denudation.  Such a situation arises when a
basin encroaches on the adjacent dry land.  Such conditions existed during
the formation of the Gulf Coast oil-gas basin, each new oil-bearing formation being connected with a successive advance of the sea.
    The coastal zone is, therefore, characterized not only by the intensive
accumulation of organic matter in the sediments of the lagoons, bays, limans,
deltaic interfluvial water bodies, etc. , but also by its rapid burial beneath
the sediments of the advancing sea.  This is why the organic matter escapes

oxidation.  A regressing sea usually leaves behind a broad strip of relict lakes and marshes along the coast; these water bodies also serve as sites of accumulation of abundant organic matter of the humus or sapropel type. The alternation of transgressions (ingression) and regressions leads to an interfingering of marine and continental facies.  This explains why oil- and gas-bearing deposits are often associated with zones of transition from continental subaqueous sediments to marine sediments, and vice versa.  This transition is often marked by the presence of oil-, gas- and coal-bearing formations in the section of one and the same stratigraphic subdivision.

A favorable combination of all the conditions necessary for the accumulation and burial of organic matter and its subsequent transformation into hydrocarbons, as well as for the formation of oil pools, is encountered in coastal zones with deltaic deposits.

*River Mouths: Influence on Occurrence of*
*Oil and Gas in the Coastal Zone*

When speaking of the association of oil fields with deltaic deposits, we have in mind formations of the delta proper, its underwater portion — the prodelta, and other adjacent accretion forms.  One must bear in mind that each river mouth has its own specific features.  Delta formation is influenced by both the action of the river and by marine hydrodynamics.  If the sea is the predominant agent at the river mouth, the waves can carry sediments into it and divert its course by tens or hundreds of kilometers. Strongly flowing rivers usually produce bars or subaqueous spits fringing the extension of the distributaries or the estuarine projection.

In cases where the fluvial and tidal sections were repeatedly shifted during the process of delta growth, due to the redistribution of the flow along various distributaries, the resulting delta will have a very complicated structure.  Many rivers may have flowed through vast shallow-water limans or lagoons with muddy bottom sediments before entering the sea. This is why coarse alluvial material is only encountered in ancient distributaries and near the mouth in the form of river bars.

In places where the river mouth was actively influenced by surf the sand deposits form a series of natural levees, each of which marks the position of the shoreline at a different stage of delta growth.  The simplest alluvial projection of the delta with a shore of the surf type may form a single-channel mouth.  The river builds up a localized extension which comes under greater wave action the further it advances into the sea.  As a result, most of the sand material brought by the river is carried away from the mouth and is used for the formation of two lateral spits or bars.  Such bars are rather common for many present-day rivers characterized by marked advance.  The water bodies beyond the lateral bars develop into lagoons.

Surf-type deltas sometimes have several large distributaries; a projection is built in front of each of these channels.  If the rates of their accretion are sufficiently high, the areas between the mouths of the distributaries are not filled with coarse material and form indentations as in the Nile Delta.

Experience has shown that the detailed character of many paleodeltas, which contain abundant reserves of oil and gas, can be reconstructed using the comparative paleogeographic method.

We shall discuss some other examples, cited by Veber (1956), confirming the association of oil fields with deltaic or near-delta formations.

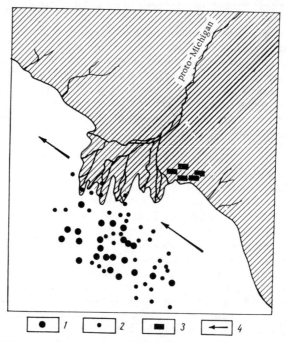

FIGURE 27. Paleogeographic reconstruction of the Illinois basin during the Late Mississippian (according to Swann with additions by Veber, 1966):

1 — oil fields from which more than 1.5 million tons of petroleum were produced; 2 — fields from which less than 1.5 million tons of oil were produced; 3 — coal; 4 — direction of littoral currents.

Figure 27 shows a paleogeographic map of the Illinois basin during the Late Mississippian. Oil and coal fields occurring in the same stratigraphic interval — Chester — are designated. Despite all the shifts of the shoreline and the delta of the proto-Michigan, the coincidence of oil fields and deltaic deposits is beyond doubt.

Oil pools at the mouth of this ancient river occur in thick sand bodies which were deposited, according to Swann (1964), mainly seaward of the delta distributaries and to a lesser extent in the delta distributaries proper. These sand bodies are genetically comparable with bar-finger sands. Another portion of the estuarine sands is described as lunate sand bars. The prodelta sands are thinner and of smaller extent; they were formed as a result of redeposition of estuarine sand. All sand varieties alternate with argillaceous deposits and limestones.

Another example is the Tertiary delta of the paleo-Irrawaddy in Burma. This river once flowed into the northern part of the Bay of Burma. Later the delta shifted southward and the sea receded from the bay, in which rivers

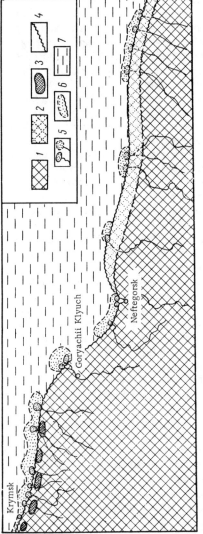

FIGURE 28.  Paleohydrographic and facies map of the southern part of the Kuban depression at the end of the Early Maikop (Maslov, 1962):

1 — dry land; 2 — assumed dry land; 3 — coastal promontories; 4 — shoreline; 5 — paleorivers and their deltas; 6 — sand zones of deltas and prodeltas; 7 — region of accumulation of argillaceous sediments.

and ephemeral streams deposited the Irrawaddy formation of continental facies. Petroleum occurrences in Burma are associated with sedimentary deposits of the shallow sea and estuaries fringing the outer part of the delta. Oil here is believed to be syngenetic with the enclosing rocks.

Maslov (1962) detected and described a number of oil-bearing deltas of Early Maikop age in Northern cis-Caucasus, on the southern flank of the Kuban depression. In the Neftegorsk-Khadyzhenskaya region he discerned the lower courses of the paleo-Tukh, paleo-Pshish, and paleo-Pshekh river valleys. Here I. M. Gubkin discovered the well-known producing channel sandstones. While tracing these sandstones, Maslov was able to locate the site where the ancient river debouched into the marine basin. He also established the presence of the buried paleo-Psekups valley which passes through Goryachi Klyuch, following a course approximately coinciding with that of the present-day Psekups valley.

The paleohydrographic map (Figure 28) shows the sand deposits occurring in the deltas and prodeltas of paleorivers. The modern drainage network in this region shows signs of being inherited from the Early Maikop hydrographic network. The paleorivers supplied the marine basin with sand and silt material brought from the Caucasus dry-land region where ancient sedimentary strata were undergoing erosion. The erosion products were deposited in valleys, deltas and prodeltas, as well as in adjacent nearshore portions of the ancient basin. There is no evidence of the terrigenous material being carried by the currents over large distances.

The map also shows the most probable location of the deltas and prodeltas of the Belaya, Khodzha, Laba, Urup, Zelenchuk and Kuban paleorivers, as well as the associated sand zones. The zone of sand formations in the eastern part of the ancient coast possibly falls within the boundaries marked on the map, provided the shoreline did not pass to the south of the second tentative line. The width of this strip, according to Maslov, does not exceed that of the sand zone associated with the deltas of the paleo-Pshish and paleo-Pshekh, i. e., the zone which has been covered by drilling. He rightly concludes that the deposits at the mouths of the above paleorivers, including the adjacent sand zones, may well be "happy hunting grounds" in which lithologic pools may be discovered. It is known that several channel sand bodies have already been traced within the paleo-Pshekh prodelta. These contain the main oil pools. In this region oil has also been discovered in sand lenses which are widely distributed along the distributaries of the paleoriver.

Veber (1966) noted that the sand bodies of the paleo-Pshekh prodelta closely resemble the bar-finger sands of the Mississippi delta, which are formed in the estuarine portions of the deltaic distributaries, as described by Fisk (1961). A zone of "pure sand," made up of fine sand and coarse silt, is encountered in the central portions of such sand bodies. Silt predominates in the marginal parts of these sand bodies. They are fringed and underlain by clayey silt of the delta front (Figure 29). Silty clays and clays of the prodelta are deposited further seaward.

Fisk considers the Booch sand of Lower Pennsylvanian age in the Greater Seminole district, Oklahoma, to be analogous to ancient bar-finger delta sands with which oil pools are associated.

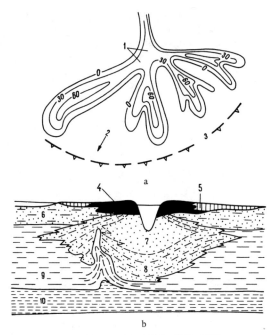

FIGURE 29.  Bar-finger sands of the Mississippi delta (Fisk, 1961):

a — plan and thickness of sand body (in m); b — cross section.  1 — upstream tapering of "fingers"; 2 — seaward widening of "interfinger" areas; 3 — edge of delta platform, transition to depths greater than 73 m; 4 — natural levee; 5 — marshes; 6 — delta plain; 7 — "pure" sand zone; 8 — transition zone; 9 — delta front; 10 — prodelta.

## Liberation of Hydrogen in the Coastal Zone

In addition to the above-mentioned conditions favoring the formation and accumulation of hydrocarbons in the coastal zone, we shall mention yet another phenomenon which may be considered as a factor contributing to the increased productivity of this zone.

It is common knowledge that the presence of free hydrogen is a prerequisite for transformation of organic matter into hydrocarbons.  A possible source of hydrogen is via the process of superfine comminution of rocks and minerals in water.

Golosov, Dolgov, Molchanov and Shugurova (1966) (research workers of the Institute of Geology and Geophysics, Siberian Branch of the U. S. S. R. Academy of Sciences) established experimentally that hydrogen is produced during the process of comminution of minerals in an aqueous medium.  The liberation of hydrogen is due to the reaction between water and suspensions of this comminuted mineral material.  The reaction continues for a long period of time and is intensified by mechanical stirring because the quantity of fine particles increases on disintegration of larger aggregates.

The above-mentioned scientists stress that this mode of generating hydrogen is totally independent of the gases contained in mineral inclusions.

The pure quartz used in their experiments was free of gas inclusions. Microchemical analyses of individual gas inclusions in honeycomb quartz revealed the total absence of hydrogen at varying contents of $H_2S$ and $CO_2$.

It is assumed that the liberation of hydrogen is due to: 1) dissociation of water by finely dispersed iron; 2) formation of hydrosol micelles; 3) formation of new minerals. In the last case, for instance, on comminution of orthoclase one observes the entire course of weathering with alteration to muscovite and kaolinite.

The pulverization of minerals in an aqueous medium leads to their leaching. The capacity for cation exchange increases: $K^+$ is replaced by the $H_3O^+$ (hydroxonium ions). Potassium in solution reacts with water and silicic acid. The following is a possible reaction giving rise to the liberation of free hydrogen:

$$2H_2SiO_3 + 2KOH \rightarrow 2KSiO_3 + 2H_2O + H_2.$$

Other alternatives are possible, but we shall not discuss them here. Fine comminution of minerals in an alkaline medium may lead to the formation of clay minerals of the montmorillonite type, the outer layers of the structure of which consist of the hydroxyl group $OH^-$. The removal of the latter from the water is accompanied by the liberation of hydrogen ions.

These were first attempts at explaining the chemical transformations leading to the liberation of free hydrogen during superfine comminution of mineral particles in an aqueous medium. This process should be studied more carefully both under experimental and natural conditions. According to recent communications,* Molchanov and Gontsov, who checked their assumptions in the laboratory, succeeded in deriving an artificial petroleum-like mixture of hydrocarbons by hydrogenating organic material under conditions close to those prevailing in nature.

In any event, under natural conditions this phenomenon could only occur in the coastal zone to which, firstly, huge amounts of dispersed material are transported and where, secondly, this material is subjected to prolonged wave action. We have focused attention on this aspect not only because it may shed new light on the reasons for the higher productivity of oil and gas fields in ancient coastal zones, but also in order to arouse the interest of prospective researchers. The experimental data must yet be checked in natural environments. If they are eventually confirmed, the paleogeographic principles of detecting rich oil- and gas-producing areas will gain additional support.

It is possible that the interaction between the forces of wave motion and of littoral currents in the coastal zone, especially during stormy periods, leads to extremely vigorous stirring of the superfine mineral particles, the likes of which cannot be reproduced experimentally. Moreover, the time factor under natural conditions probably compensates for the brief duration of hydrogen liberation under experimental conditions.

* "Tyumenskii Geolog," 1970, No.25 (311).

## 7.  THE SHELF

The sea bottom covered by shallow waters fringing the continents is known as the continental shelf.  The shelf is usually bounded by a sharp flexure in the bottom section which marks the transition to the continental slope and delineates the outer margin of the shelf.  Until recently it was believed that the continental  shelf extends to a depth of 200 m.  However, studies of the shelf relief have shown that it may stretch to even greater depths.  Therefore, those parts of the sea bottom which retain the character of the shelf relief are called the "subaqueous continental margin" (Leont'ev, 1963).

Shallow seas with depths characteristic of subaqueous continental margins are referred to as shelf seas.  Such seas may either be marginal, i. e., located at the continental margin and having relatively free access to the ocean, or else they may be inland and are connected to marginal seas via narrow straits.

The most important feature of inland seas is the fact that more clastic material is transported into them than into marginal seas.  Moreover, their waters may have different salinities.  Shallow seas, such as the Sea of Azov, the Baltic Sea and the White Sea, are similar to inland seas.  Being part of the inundated continental platform, they have a relatively gentle bottom relief and bear all features of inland shelf seas.

Large lacustrine basins lacking outlets to the ocean constitute a separate group.  These include the Caspian Sea, the Aral Sea, Lake Baikal, etc.  The first two are classified as seas by reason of their geological past.  Although the sedimentary environment of inland seas is similar to that of shelves, the process of sedimentation is more diverse in the former, being dependent not only on the basin depth, but also on river runoff, the size of the water body and other factors.

In this work the concept of shelf includes shallow seas.  In recent years geologists have been paying ever increasing attention to the continental shelves, especially in connection with the search for oil and gas pools.  The factual material pertaining to present-day shelves is rather sparse and even less is known about ancient shelves.

Studies of modern shelf zones of different seas make it possible to classify them according to several principal criteria.  One might regard uplifts and depressions of the earth's crust, with which the development of shelves is connected, to be an important criterion.  However, the division of shelves into "uplifted" and "downwarped" is a difficult task, especially in view of the effect of regional glaciation on the level of the ocean.

**Shelves fringing regions of glaciation** are usually wide and have a dissected relief.  They contain numerous basins and troughs whose depths normally exceed 200 m.  The shelf depressions may extend both parallel and perpendicular to the strike.  Numerous sand banks stretch along the outer margin of the shelf.  Silty sediments with a considerable admixture of coarse clastic material fill the basins and troughs.

**Shelves with elongated sand banks and hollows.**  The majority of shelves located near coasts unaffected by glaciation have a flatter surface than that typical of the preceding type.  However, even these shelves are not straight and are characterized by a convexo-concave bending of the slope that extends continuously from the shore to the shelf margin. Other irregularities include long sand banks or ridges which run parallel to the shore in the inner part of the shelf and along the shelf bend at the outer margin. Shallow depressions or hollows are often encountered between the banks.

**Flat shelves in the proximity of coasts in regions of higher latitude.** The available bathymetric data indicate that the flattest continental shelves are those in northern latitudes, along shores that had not experienced glaciation. Such shelves show insignificant surface irregularities due to the action of the bottom ice. Shelves of this type usually occur at smaller depths than those average for the shelf zone.

**Shelves associated with the action of strong currents.** Permanent currents run along many coasts. It is noteworthy that in most cases zones with the strongest currents coincide with regions with very narrow shelves or totally lacking shelves. Most of these narrow shelves either have a rocky bottom or are covered with sand or gravel transported by the currents. Strong currents in the narrow mouths of bays (straits) form deep basins.

**Shelves in front of deltas of large rivers.** Such shelves are usually very wide, notwithstanding the fact that parts of their territories are occupied by the growing deltas. Nearly all shelves in front of the mouths of large rivers are covered by muddy sediments, usually at the edges of deltas, whereas sands are more common on the outer side of shelves. Mud deposits along the shoreline contain large amounts of plant fragments and detritus, mica and ferruginous formations.

**Shelves of tropical seas.** These shelves are distinctive in that coral or algal reefs rise above the shelf bottom in those localities where the sea water is not made turbid by debouching rivers. The reefs generally tend to occur at the outer margin of the shelf.

These are the general geomorphological features of the main types of modern shelves. The same features, with certain modifications, are also characteristic of ancient shelf zones.

How much organic matter is contained in the shelf deposits? An idea of this content may be gained from the fossilization coefficients for organic matter in various parts of oceanic facies (in %): open ocean (depth of 2450 m) — 0.6; continental slope (depth from 200 to 2450 m) — 0.37; shelf sea regions (depth of 200 m) — 1.04 (Uspenskii, 1970).

The next links in this chain are the lacustrine (fossilization coefficients from 3 to 5%) and palustrine facies (average coefficient 8.6%). However, in most cases the organic matter in continental facies serves as the source material for coal accumulation, whereas that in marine facies is mainly the starting material for oil formation.

The distribution of organic matter is known to bear a direct relationship to the lithofacies of the rocks, the higher concentrations generally being observed in argillaceous rocks. The amount of organic matter in carbonate rocks, as well as in sandstones, depends on their content of clay material.

Even though the organic content in rocks is not, in itself, indicative of the actual possibilities of oil formation, the fact that the majority of oil pools are encountered in broad shallow-water zones of ancient seas reveals that the shelf regions are in all respects favorable for oil and gas accumulation. Regional oil and gas occurrences of many modern shelves can apparently be attributed to the relatively continuous and prolonged formation of the shelf zone within the same downwarped portion of the earth's crust.

There are differences of opinion on the origin of continental shelves.
F. P. Shepard, for instance, considers the formation of most terraces on the
outer portions of shelves to be the result of abrasion at a lower sea level,
or the outcome of delta formation at a time when the water level in the ocean
was low. He believes that several wide shelves are only partly due to fluctuations in the sea level, caused by glaciation. Some of them could have been
formed by sedimentary infilling of relatively deep troughs. Shelves similar
to those off the east coast of the U. S. A. have undergone a prolonged stage of
subsidence and accumulation of sedimentary material, the latter phenomenon
being in many cases connected with delta formation. It is possible that many
wide shelves are formed as the result of downwarping and subsidence of the
continental margins, leading to their sinking below sea level, partly as the
result of the rock overburden and isostatic adjustment.

Fairly broad shelves could originate in regions where alluvial plains have
subsided as the result of tectonic movements. This could lead to the formation of shelves only in those cases where the rate of subsidence of such dry-
land areas exceeds the rate of sedimentation.

The average width of the continental shelf is about 75 km. The average
depth of the shelf edge (where the sea bottom bends most sharply) is 130 m.
Knolls, 20 m or more in height, are detected on 60% of the bottom profiles.
Depressions with a depth exceeding 20 m are encountered on 35% of such
profiles. Many of these have the shape of closed basins, while others form
elongated hollows.

The concept that the sediments on the shelf are well sorted and range
from coarse-grained near the shore to fine-grained at the outer margin of
the shelf is outdated. Analyses of samples of shelf sediments have revealed
that the distribution of the sedimentary zones does not conform to any particular pattern and is little related to either the distance of a certain portion
of the shelf from the shore or to the shelf depth.

In this context it is of interest to refer to one of the broadest shelves off
the coast of China. It bears the closest resemblance to the ancient shelf
seas. Rivers, such as the Hwang Ho, Yangtze and Mekong, supply sedimentary
material to the shelves of the East Asian coast. Mud deposits predominate along
the shores, and sands on the outer margins of the shelves. Japanese and
American investigators have found that within the western part of the East
China Sea and in the South China Sea the sediments are mainly fine-grained,
consisting of redeposited loess, while in the east they are more coarse-
grained.

North of the Irrawaddy delta, along the Burma coast, the shelf narrows
down to 37 km, the sediments consisting mainly of mud. Near the mouth of
the Ganges the shelf is 185 km wide and at a depth of about 110 m. The shelf
narrows down to 110 km due to the advance of the Ganges delta.

The inner part of the Gulf of Oman has a broad shelf which extends into
the shallow-water zone of the Persian Gulf. Its bottom is covered with mud
brought by the Tigris and Euphrates rivers. These two rivers form a common delta, Shatt al Arab, which, according to Emery (1956), has advanced
165 km into the Gulf over the last 4000 years. Sands predominate in the
southern and western parts of the Persian Gulf. Emery noted the rather high
content of organic matter in sediments of the Gulf's outer portion, where, in

the Strait of Hormuz, there are concentrations of plankton resulting from
the mixing of the waters of the Arabian Sea and the Persian Gulf.

It follows from the above examples that the bottom sediments of continental shelves, even in regions where large rivers debouch into the sea,
generally consist of muddy material. Coarse-grained sediments usually
occur near the coast, within the intradeltas, sometimes also on the outer
margin of the shelf. It is very important to bear in mind that the distribution of sediments on open shelves differs from that in inland shelf seas. This
difference is determined by the bottom relief, i.e., the character of the base
prior to marine transgression, the strength of the currents, river runoff and
the depth of the basin.

Deviations from the general regularities of the shelf structure have certainly occurred in the past, and they also exist at present. They result from
the combined effect of various factors which must be taken into account during reconstruction of the paleogeographic environment of one of the most
interesting regions of sedimentation and formation of oil and gas pools.

### Features of Shelf Deposits

The paleogeographic environments of sedimentation on ancient shelves
are usually established by analogy with present-day shelf formations. However, the majority of the ancient terrigenous rocks must have been deposited
in closed epicontinental basins, the so-called shelf seas, where depositional
conditions differed from those on present-day continental shelves. Therefore,
when one studies producing formations of the shelf zone, one should pay attention to the diverse forms of distribution of the sand reservoirs.

Shepard (1969) distinguishes the following types of ancient sand deposits:
sheet sands, shoestring sands and deep-water sands. We shall discuss them
in some detail.

**Sheet sands.** These ancient marine sands sometimes cover enormous expanses, for instance, the St. Peter marine sands, etc. We are puzzled by the
fact that the sands in ancient seas were apparently transported over such
considerable distances, whereas the waves of present-day seas do not seem
to be able to carry sands for more than a few kilometers away from the
shore. There is the possibility that slowly transgressing seas in different
periods of the geological past reworked ancient alluvial sands of nonmarine
origin, as the shoreline shifted. On gently sloping areas of the subsiding
sea bottom the alternation of transgressions and regressions led to the appearance of a corresponding amount of transgressive and regressive sandstones of enormous lateral extent.

According to Shepard, tidal action leads to another sand formation which
also has quite impressive dimensions. It may be encountered both on the
present-day sea bottom and in littoral shoals. For example, the sand plain
near the Bay of Mont-Saint-Michel (west coast of France), which was built
by tidal currents, is about 13 km wide. This plain consists almost entirely
of sand. The shoals have well-developed ripple marks with transverse
scour marks. Sheet sands form the southern extremity of the North Sea,
where they have a very uneven surface.

Shoestring sands were discovered in a number of regions of the U.S.A. during the drilling of wildcat wells. These are generally long, narrow sandstone lenses occurring among argillaceous rocks. In places they extend for tens of kilometers. Geologists believe that some of these lenses are offshore sand bars and that others are channel fillings. Offshore bars usually have a straight or gently curving frontal part facing the sea and a dissected, meandering part adjacent to the lagoon. Channel fillings are completely different from offshore bars and usually have parallel meandering outlines. Sediments of offshore bars and channel fillings also differ markedly in their grain size and degree of sorting. In offshore bars well-sorted sand occurs on the seaward side, while on the side transitional to lagoon deposits the sediments are markedly silty. Channel-filling deposits are often poorly sorted owing to the repeated changes in the meandering course of the river; interbedding of argillaceous varieties and sandstone lenses is not uncommon.

Deep-water sands are encountered among thick shales which were, in all probability, deposited in deep, calm waters. In ancient deposits, especially those of geosynclinal formations, they are represented, to a significant degree, by coarse-grained or even conglomerate-like layers or lenses of sandstone. It is believed that these sands were transported by turbidity currents which flowed down submarine valleys located in the marginal parts of ancient basins; the deep-water basins of California can be regarded as their present-day analogues.

These deposits are not readily recognizable. They are usually characterized by graded bedding, each sand layer consisting of coarser particles in its lower part and finer particles in its upper part. These sands generally contain considerable admixtures of silt and clay particles. Many of them show coarse- and fine-grained layers. It is rather difficult to trace the source of the coarse-grained material. Both deep-water and shallow-water sands are made up of virtually the same material. Sometimes one encounters micas and plant fragments. The fauna in sands of turbidity currents may consist of a mixture of shallow-water and deep-water species.

Among other marine deposits described by Shepard we should like to mention clay deposits, which are the most widespread variety of ancient sedimentary formations.

Mud deposits on the shelf cover large expanses and on present-day shelves are located in front of the estuaries of large rivers. Examples are the vast mud zones of the inner shelf off the east coast of Asia and areas to the east and west of the Mississippi delta. The bedding of the present-day shelf muds is due to the seasonal nature and vital activity of the benthic organisms. Locally, the mud contains isolated sand lenses with large quantities of planktonic foraminifera.

Lagoon mud predominates in large lagoons situated inside many barrier beaches. It differs somewhat from the mud deposits on open shelves. Glauconite and fossil sea urchins are rarely encountered in lagoon mud. Gypsum deposits occur in lagoons of semiarid regions and salt beds in those of arid regions. Ancient lagoon formations are usually fairly narrow and oriented along the shoreline. However, shoreline migration due to transgressions and regressions could explain the large areas occupied by lagoon deposits. The fauna of landlocked lagoons and bays is considerably poorer in species than the shelf fauna.

**Delta front muds.**  River muds are in part deposited in the floodplain, but are mostly carried into the sea.  In places where the mouths of lowland rivers approach the sea, in the absence of tides or high waves, mud deposits of considerable extent are built up in the marginal parts of the delta.  In the small bays of delta distributaries relatively thick layers of mud alternate with much thinner intercalations of coarse silt or fine sand.  The bedding of the mud and its interlayering with sand material usually fade out downward from the top of the delta slope.  Absence of bedding is characteristic of muds located beyond the delta slope.  Delta deposits also have a higher content of mica than shelf sediments which accumulated at a distance from the river mouths.

The above are the most important features of certain nearshore and shallow-water marine deposits with which commercial deposits of oil and gas are frequently associated.  Although these features in most cases relate to recent or fairly recent sediments, they should also be typical of similar ancient physiographic environments of sedimentation.

Muddy sediments of shelves, lagoons and deltas, as mentioned above, are the materials most prone to subsequent transformation into source beds.  If such mud deposits are located in close proximity to sand deposits possessing good reservoir qualities, this may create favorable conditions for vast zones of oil and gas accumulation.

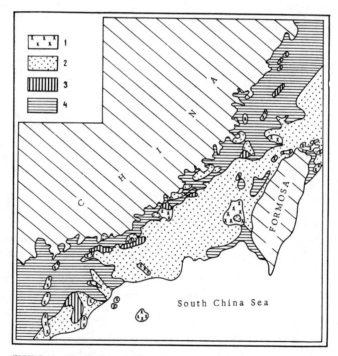

FIGURE 30.  Distribution of sediments on the shelf:

1 — bedrock;  2 — sand;  3 — silty sand;  4 — mud.

There are a number of cases where the transition from sand to clay formations on the shelf does not coincide with transition from the coastal to
the deep-water zone.  For instance, on the present-day shelf of southeast
China (Figure 30) the sand deposits are situated far from the shore, beyond
the belt of mud.  A similar distribution of shelf formations can be observed
along the west coast of Africa, as well as in the Caspian Sea.  The abundance
of muds on the nearshore part of the shelf is due to their intensive transport by rivers.

## 8.   HIATUSES AND UNCONFORMITIES

Hiatuses and unconformities are one of the connecting links between the
tectonic regime and the paleogeographic environments of sedimentation.
Although they are of geotectonic origin, they do reflect the influence of the
physiographic conditions.  Most hiatuses are known to have occurred on
continents and are much less frequent on the bottom of marine basins.  Prolonged regional hiatuses can be traced over vast areas of the crust.  They
are usually accompanied by weathering and erosion of the previously deposited rocks or by a temporary break in sedimentation.  Surfaces of erosion marking the hiatuses are of great significance in the reconstruction of
ancient physiographic environments.
    In addition to regional hiatuses, there are also local hiatuses and degra-
dation vacuities connected with the tectonic structure and the action of
rivers, ephemeral streams and undercurrents.  Short breaks in sedimentation may not be accompanied by erosion of the older deposits.  In such
cases the hiatus surface bears traces of the ancient landscape: ripple marks,
desiccation cracks, soil layers with remains of plant roots, etc.
    The appearance of a weathering crust is a distinct sign of a prolonged
stratigraphic break.  It represents residual products of the disintegration
and chemical alteration of metamorphic and igneous rocks.  For instance,
remains of a lateritic weathering crust are indicative of a planar relief and
a hot and humid climate, while a red crust suggests a hot and intermittently
humid climate.  Often one encounters, together with the weathering crust,
altered sedimentary rocks whose composition is similar to that of the rocks
of the weathering crust.
    The tectonic regime is the main factor controlling the formation of a
weathering crust (Kazarinov, Bgatov, Turova et al., 1969).  The climatic
factor, which either speeds up or slows down the chemical decomposition
of rocks, determines the type of weathering crust.  The latter does not include redeposited products, such as quartz and quartz-kaolin sands, refractory clay, and other sediments located at a distance from residual formations.  Redeposited terrigenous and chemical products of the weathering
crust generally accumulate in regressing basins during the initial stages
of orogeny.
    The intensity of weathering is dependent both on the relief and the climate.
The smoother the relief, the weaker the development of the weathering crust.
Conversely, this process is particularly active during the uplifting of the
large blocks of a peneplain under conditions of marine regression.

Deposits of the weathering crust are subdivided into residual and sedimentary. The former are represented by eluvium, one variety of which re-
tains certain structural features of the bedrock. The latter are redeposited
products of distant transport at the time of formation of the eluvium with
residual structure. Chemogenic sediments sometimes account for much of
their composition.

The above authors believe that a characteristic feature of these sediments
is the presence of slightly altered or even fresh unstable minerals among
some redeposited products of the weathering crust. The accumulation of
these minerals ends with the complete or almost complete cessation of
weathering crust formation on dry land, when the overall rate of the land's
uplifting exceeds that at which the eluvium is formed. This is the reason
why chemogenic sediments are found in the redeposited products of the
weathering crust. The siliceous formations of the West Siberian plate
mantle are a vivid example of such formations.

One may gain an idea of the relief of a region undergoing erosion by
studying the grain size of the clastic rocks in the adjacent regions; sometimes the mineral composition of the deposits is also informative. Coarse-
grained material is supplied from source regions with highly dissected re-
liefs. A considerable amount of products of chemical weathering are derived
from a planar relief. The size of pebbles in fluviatile conglomerates is
sometimes indicative of the relative height of the denudation region. The
rate of uplift of such areas is reflected by the granulometric composition of
sediments, the rivers in this case having a greater erosive effect and carrying coarse-grained and poorly sorted material. If the rate of uplift of the
region undergoing erosion is slow, the terrigenous material is generally
fine-grained and well-sorted. Atmospheric precipitation, i.e., the climatic
factor, also has an important effect on the activity of rivers.

Consequently, breaks in sedimentation prior to marine transgressions
are sometimes expressed as missing parts of the normal stratigraphic section, sometimes as traces of erosion and redeposition of rocks, or, at the
margins of a basin, as weathering crusts. Hiatuses may play both positive
and negative roles in the accumulation of oil and gas.

One may obtain a general idea of the paleorelief resulting from a hiatus
on a given territory from an isopach map of the sediments covering this
relief. A reduction in the thickness of the post-hiatus strata corresponds
to positive relief forms, while an increase in their thickness corresponds
to negative forms.

The following serves as indications of breaks in marine sedimentation
(Yablokov, 1971):

angular unconformity relative to underlying rocks;

erosion surface, usually uneven, undulating, truncating underlying deposits
at various depths;

weathering crusts of different thickness and structure;

karst formation and weathering of carbonate, argillaceous, and arenaceous
rocks in individual horizons;

bedding planes with desiccation cracks and breccias;

intercalations of rocks with traces of autochthonous root systems of
plants, paleosols;

autochthonous coal beds and arenaceous-argillaceous rocks of lacustrine origin;

    intercalations of pebbles and conglomerates;

    paleovalleys and arenaceous rocks of alluvial origin;

    rocks of eolian origin;

    argillaceous-arenaceous glacial boulders;

    abrupt change in fauna not attributable to facies changes in the given area.

The above-enumerated features are not necessarily indicative of a continental hiatus. For instance, angular unconformities may be associated with cosedimentation structures at the bottom of the basin; minor erosion may be due to undercurrents; pebbles and conglomerates may be of both continental and marine origin.

Hiatuses detected in individual cross sections should be traced areally with the aid of the above-listed criteria. One should also have an idea of the depositional environments prior and subsequent to the presumed hiatus. This will make it possible to differentiate between the erosion surfaces associated with intraformational hiatuses and those associated with stratigraphic breaks.

It is a much more difficult task to assess the duration of continental hiatuses. Yablokov suggests several approaches to the solution of this problem, but all of them require further development. Some of them are listed below.

1. According to the difference in absolute age of the rocks immediately underlying and overlying the erosion surface. It should be borne in mind that the hiatus may in fact be of somewhat shorter duration, as younger deposits may have been destroyed at the beginning of the hiatus.

2. According to the overall duration of a given geological period. For instance, in the central regions of the East European Platform deposits of the Givetian stage (Middle Devonian) overlie Ordovician rocks. Consequently, Silurian rocks, as well as those of the Lower and part of the Middle Devonian were not deposited. The total duration of the hiatus in this case is 50—55 million years.

Another example is the pre-Radaevsky hiatus in the Lower Carboniferous. The Upper Tournaisian Chernyshinsky limestones are overlain by the eroded arenaceous-argillaceous Radaevsky horizon. Rocks of the Kizel and Elkhovo horizons, whose deposition period in other regions is estimated at 5—6 million years, are missing. Therefore, this time gap is assumed to be the duration of the hiatus in this case.

3. According to the rate of rock denudation. For this one requires data on thickness of the rocks destroyed during the hiatus and on the rate of denudation and erosion. The thickness of the eroded strata is assessed by analysis of the corresponding interval of the stratigraphic sections over large areas. The indexes of erosion are determined for various climatic conditions and various rocks.

4. According to the time of accumulation of the synchronous strata in adjacent areas. This calls for knowledge of the rate of accumulation of the series and of their thickness. Crustal movements in the given area and the hydrodynamics of the basin should be taken into account when calculating the rate of sedimentation.

The above methods of determining the duration of an hiatus are somewhat arbitrary, but they can nevertheless help us to determine at least the relative, if not the absolute, duration of the break in sedimentation on a given territory.

Very important for paleogeographic reconstructions is the determination of the duration of the continental conditions of sedimentation when terrestrial formations accumulated on surfaces exposed following a marine regression. We employ the same principle as applied for assessing the duration of hiatuses, the difference being that we are interested in the period of time necessary for the accumulation of material, and not its erosion.

Even though regional gaps in sedimentation are caused by epeirogenic oscillatory movements of the earth's crust, they are influenced by physicochemical processes determined by the paleogeographic environment. It is common knowledge that the following conditions are necessary for the formation of oil and gas: relatively prolonged and steady subsidence of the sedimentary basin and burial of the sediments containing organic matter at optimal depths. Hiatuses accompanied by erosion naturally disturb sedimentation and sometimes bring about the destruction and erosion of rather large bodies of previously deposited rocks. They could also lead to the destruction of already formed oil and gas pools, this being the negative effect of an hiatus on oil and gas accumulation.

However, such breaks may also favor the generation of reservoir rocks and paths of migration for oil and gas. The important point here is that the processes of weathering during an hiatus create an erosion surface below which a zone of leaching is formed. In such cases the rocks disintegrate or become permeable to groundwaters, i. e., they become porous, cavernous and jointed.

Uspenskaya (1952) has shown that on the North American Platform there is not a single large oil- or gas-bearing horizon in limestones, which is not associated with an erosion surface. About 95% of the entire production from carbonate reservoirs in the U. S. A. is from horizons underlying unconformities. The following are examples of the direct relationship between the productivity of carbonate reservoirs and unconformities: the oil and gas pools in the Ordovician limestone-dolomite formations in the Lima-Indiana region, in the Devonian limestones of the Michigan and Eastern Interior basins, in the Devonian, Mississippian and Ordovician limestones of the Western Interior basin, and those in the Permian limestones and dolomites of the Permian basin.

Areas saturated with petroleum show a zonal distribution in the above carbonate reservoirs. They are associated with individual porous and permeable, in places cavernous, zones resulting from solution and leaching of carbonates by groundwaters or perhaps from their dolomitization. These zones are located at varying depths below the erosion surface, ranging from several to hundreds of meters. For instance, the main producing zone in the Ordovician limestones of the Oklahoma City field is 150 m below the principal unconformity surface, which is overlain by Pennsylvanian beds.

Uspenskaya points out that porous oil-bearing zones in limestones are generally of local occurrence and are associated with the arched parts of structural uplifts. Sometimes they occur in buried projections of the ancient relief composed of carbonate rocks. In rare cases oil and gas accumulations occur in weathered and jointed crystalline rocks forming projections which are unconformably overlain by sedimentary deposits. A comparison of the general character of lithogenesis in the North American and East

European platforms shows that the major producing horizons of the Volga-Ural region, like those in a number of oil and gas basins of the U.S.A., occur in the upper horizons of pre-hiatus deposits.  In particular, hiatuses have been detected in the Upper Devonian carbonate beds occurring between salt beds in the Rechitsa oil field.

Angular unconformities, with impermeable beds deposited after the hiatus overlying tilted strata of porous rock, are another example of the favorable influence of hiatuses on oil and gas accumulation.  These porous rocks may serve as reservoirs, while the impermeable beds may seal off the pools.

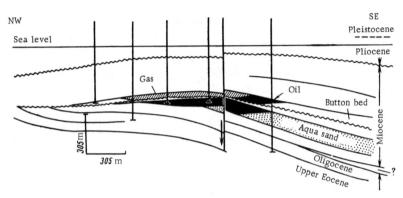

FIGURE 31.  Section through the Antelope Hills oil field, California (Levorsen, 1970)

In the case of faulting oil pools may occur both above and below the unconformity surface (Figure 31).  One of the largest oil fields in South America — Quirequire — is located in the Maturin basin.  Here the producing formation, consisting of unsorted clastic rocks ranging from silty sands to boulder beds, overlies an erosion surface of folded and faulted Eocene-Miocene strata.

Zones of truncation and onlap, where stratigraphic traps may be formed, as well as facies pinch-out zones, usually occur on the slopes of uplifts or along the flanks of depressions.  A closer inspection of these zones reveals that they often coincide with ancient coastal zones.  Recognition of hiatuses and angular unconformities in zones where impermeable rocks truncate reservoirs occupying a favorable structural position is a prerequisite for detecting stratigraphic traps.  To this end, it is necessary to draw paleogeographic maps of individual planes of regional hiatuses and to plot on them areas of truncation of formations or, preferably, beds.  In other words, it is essential to know the age and composition of sedimentary deposits which are transgressively overlapped by other rocks.

Post-hiatus terrigenous deposits are no less important than traps associated with stratigraphic unconformities, pinch-outs, or changes in rock lithology.  The base of such deposits usually consists of coarse-grained sand which is a good reservoir.

The productivity of post-hiatus formations may be due not only to their porosity and permeability, but also to the fact that active tectonic restructuring is followed by a fairly long period of quiescence during which the sedimentary basins show a tendency toward steady subsidence.  It is during this period that the most favorable conditions for the formation of oil and gas pools arise.

Regional unconformities often serve as boundaries between different stages of geological evolution and between structural stages.  Each of them is usually characterized by specific structural features, rock composition and presence of oil and gas.  Oil and gas fields may be associated with these boundaries.  In a number of regions in the U. S. A., for instance, pools occur at the boundary between the Mississippian and Pennsylvanian, separated by a large unconformity.  Nearly all oil and gas pools discovered in pre-Pennsylvanian deposits of the Illinois, Kansas, Oklahoma, Texas and New Mexico basins are associated with structures which were formed during the same period.

Although not only rocks but also pools are destroyed during hiatuses, during the period of evolution of most sedimentary basins, notwithstanding the various stages of deformation and erosion, large commercial hydrocarbon accumulations are preserved within them.  Moreover, other conditions being equal, the larger the number of unconformities in the stratigraphic section of a given region, the greater the probability of finding commercial oil and gas deposits in that region.

Repeated fairly short breaks in sedimentation were noted in the section of the producing Mesozoic formations of West Siberia.  Yu. N. Karagodin is of the opinion that all the major oil and gas pools in that megabasin occur in formations and horizons characterized either by a short regional hiatus accompanied by erosion of the previously formed deposits or by a short hiatus in individual portions of the most mobile tectonic zones.  This is due to the fact that beneath post-hiatus beds there accumulated hydrocarbons which migrated along groups of strata and horizons truncated by erosion, rather than along a single bed.  Another favorable factor was that the regression and short break in sedimentation were followed by a new transgression, and the truncated beds were overlain by clays.

The alternation of transgressions and regressions naturally led to changes in the paleogeographic conditions of accumulation of potential source beds and reservoir rocks.  Therefore, studies of hiatuses (duration and areal extent) may be helpful not only in the reconstruction of the physiographic setting, but also in the assessment of the oil and gas prospects of a given territory.


9.   *BASIC PRINCIPLES OF PALEOGEOGRAPHIC DOCUMENTATION*

Paleogeographic documentation consists of diagrams and maps drawn to various scales, providing a fairly accurate reflection of the physiographic environments which existed in a given area during a certain period of the geologic past.  Knowledge of such environments may be gained from studies of the lithofacies, relative position and other features of rocks, both in section and in plan.

Successful reconstruction of the paleogeographic conditions of sedimentation is dependent on the amount of available factual data and on the quality of their interpretation. Much depends on the researcher's ability to "squeeze" information out of cores and well logs — direct and indirect indications of the former physiography — and apply it in paleogeographic reconstruction.

An essential condition for reconstructing the depositional environment of a formation, horizon, or member is that the material constituting the object of investigation be of the same age at all points (boreholes) along the strike. Without precise stratigraphic subdivisions and correlations of sections, it is impossible to achieve accurate paleogeographic documentation.

Yet another factor should be taken into account. The shorter the geologic time interval chosen for the reconstruction of a physiographic environment, the more reliable the reconstruction. The latter may "reflect" only a certain "moment," rather than the dynamics of historical development. It is advisable to select such "moments" within periods of maximum marine regression or transgression, when the shoreline was relatively stable. The plotting of a series of maps depicting successive changes in the paleogeographic environment is of value, as this enables one to view the process of development as if on the frames of a film.

## Paleogeographic Diagrams and Maps

It is rather difficult to make a clear-cut distinction between a diagram and a map, since even the most detailed paleogeographic map is only an approximate representation of the actual environment of the remote past. The term "diagram" is used here to indicate a tentative or very general representation of the physiographic environment of sedimentation, based on sporadic factual material. A paleogeographic map differs from a diagram in its reliability which permits utilization of the map for practical purposes. This reliability, in turn, depends not only on the amount of available factual data, but also on their accurate historical and genetic interpretation.

It is quite evident that neither the nomenclature of the paleogeographic documents, nor even in the procedure of their compilation is of essential significance here. The main point is to what extent they help to reveal the distribution patterns of oil and gas pools, and serve as a basis for prediction and the selection of the method of exploration.

The paleogeographic reconstruction of oil- and gas-bearing territories is a very complex task because of the limited information furnished by deep wells. Most works dealing with aspects of the paleogeographic reconstruction of the earth's ancient features rely on data pertaining to natural outcrops. The study of rocks in exposures obviously cannot be compared with the investigation of rocks concealed from the researcher's eye, i. e., rocks which he assesses solely on the basis of core samples or well logs. Nevertheless, the scarcity of well cuttings and core samples should not discourage geologists from attempting paleogeographic reconstructions.

The fact that these reconstructions initially have the character of fairly tentative diagrams is not of prime significance. They can later be refined and improved when more data become available. Even highly tentative but logically based conclusions broaden and deepen one's knowledge of the subject, focus attention on it, determine the future trend of studies, provoke thought and lead one to seek new ways of solving challenging problems.

Paleogeographic diagrams and maps are usually based on modern topography. They show the location of paleoseas and paleocontinents, hydrographic networks, the relief of dry land and the sea floor, and the directions of currents and winds, i. e., they reflect, as completely as possible, the physiographic environment of a given stratigraphic interval. Maps depicting only the distribution of various types of sedimentary rocks cannot be regarded as paleogeographic. It goes without saying that the reliability of paleogeographic maps depends on the completeness of the information used for their preparation. The employment of direct and indirect criteria characterizing the environments of sedimentation in the remote past will add to the scientific and practical significance of paleogeographic reconstructions.

Paleogeographic reconstruction of the depositional environments of oil- and gas-bearing strata now located at considerable depth is a rather involved task. The difficulty stems not only from the extreme paucity of core samples, which serve as a basis for paleogeographic studies, but is also due to the fact that very little is left of the former landscapes. It is especially difficult to prepare paleogeographic maps of ancient coastal zones, not to mention the delineation of the shorelines, which will be arbitrary in any event. The same, usually fragmentary information on the coastal zone, based on drilling data, may lend itself to different evaluation and interpretation by different investigators. Therefore, several variants of the same map should be prepared. It is then easier to select the optimal one. Strakhov (1970) believes that this is the only way to study such approximate entities as the elements of a paleogeographic map.

Nevertheless, even the most general paleogeographic diagrams may aid one in studying the distribution patterns of oil and gas zones and in determining the trend of exploration for oil and gas pools.

There is no need to stress the prime importance of the structural and textural features of rocks, their mineral and granulometric composition, lithofacies, physical and geochemical properties, etc. in paleogeographic reconstructions. It is beyond the scope of the present work to provide a detailed description of the methods used for studying all these characteristics. We shall focus our attention on the utilization of the results of such studies for paleogeographic documentation. However, we shall briefly discuss some of them.

### Analysis of Facies and Thickness

A most important prerequisite for paleogeographic reconstructions is a facies analysis which serves as the basis for the lithofacies maps. As stated concisely by Strakhov (1963), these lithofacies maps are "the starting point for theoretical studies," which enable one to solve problems relating to the lithogenesis and distribution patterns of oil and gas pools within a particular physiographic environment. The lithofacies map is in fact a basic document for the paleogeographer. The very first paleogeographic maps appeared simultaneously with the concept of facies in the middle of the 19th century.

Initially they depicted only the distribution of dry land and sea during individual epochs; later they reflected the evolution and alternation of physiographic environments. A great contribution was made by A. P. Karpinskii, who published maps showing the dependence of the paleogeographic environments of the East European Platform on the character of crustal oscillations.

In the course of time, facies analysis began to be supplemented by analysis of the thickness of the sedimentary deposits being studied. The method of the combined analysis of facies and thickness has been described by Belousov (1940—1954) and other authors. The application of this method assists one in reconstructing the paleorelief of individual stages of development of a given area. Moreover, isopach maps may be instrumental in revealing the distribution of reservoirs or oil- and gas-bearing formations located between two known surfaces. The more distinct these surfaces and the more reliable the determination of their depths, the more useful is the isopach map. It is worthwhile constructing such maps for producing formations or horizons of individual pools or fields and for entire regions. They may be helpful in the determination of the genetic nature of particular deposits and in studies of the geologic history of vast territories.

Isopachs can show the configuration and distribution of sand formations of river beds, bars, spits, individual lenses, reef formations and pinch-out zones, as well as cavernous or jointed carbonate horizons. The isopach map of the Booch sand (Pennsylvanian) of the Hawkins field, Oklahoma (Figure 32) and the present author's map of the Yasnaya Polyana terrigenous series in the southeast of the East European Platform (see Figure 48) may serve as examples of the distribution of productive fluviatile deposits.

Lithofacies and lithopaleogeographic maps have come into common usage in recent years. The lithology of sedimentary deposits is shown on them by line symbols, whereas the paleogeographic environments are distinguished by coloration. In the case of rock alternation the following quantitative gradation is used. If more than 90% of the rocks within the section of a particular formation (stage) are of the same type, the area over which this section is developed is marked by a single symbol. If two types of rocks occur in approximately equal proportions, each is represented by one strip of the corresponding symbol. When rocks of three types alternate, each rock type accounting for more than 40% of the entire sequence is represented by two strips.

Paleogeographic conditions of sedimentation are shown on the maps by different colors. For example, in the latest Lithopaleogeographic Atlas of the U. S. S. R., edited by Vinogradov (1969), marine basins are shown by shades ranging from dark blue to violet, depending on the depth and salinity. The dry-land features are depicted by shades from green to brown, showing the transition from coastal plains, at times inundated by the sea, to high mountains.

The same maps feature special symbols indicating groups of fauna and flora, authigenic minerals, concretions, and other characteristics of the sedimentary environment, as well as isopachs, directions of transport and marine currents, and the location of economic mineral deposits. Reef zones, ancient volcanoes and the limits of volcanic activity are also indicated by conventional signs. The reader is referred to the following sources dealing with the techniques of preparing lithopaleogeographic maps: Nalivkin (1956),

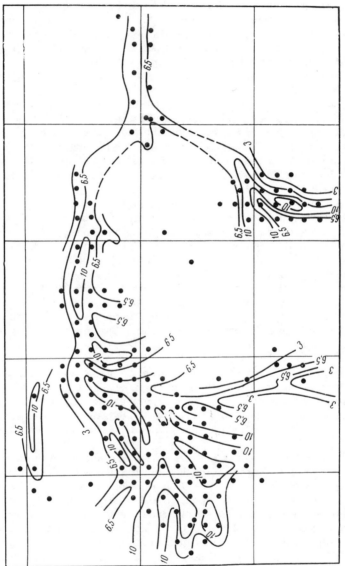

FIGURE 32. Isopach map (in meters) of the Booch sand of the Hawkins field, Oklahoma (Levorsen, 1970)

Rukhin (1959), Zhizhchenko (1959), Krasheninnikov (1960, 1962), Nalivkin, Ronov and Khain (1963), Grossgeim (1963), Mokrinskii (1963), Popov (1964), Sarkisyan and Mikhailova (1963), Einor (1963), Yablokov, Botvinkina and Feofilov (1963), Markovskii (1965, 1967, 1970), Timofeev (1970), and others.

Among the works devoted specifically to techniques of paleogeographic studies, Zhizhchenko's monograph (1959) merits special mention. It deals mainly with marine sedimentary environments of the geologic past. Much attention in this monograph is focused on the evolution of sedimentation and on methods of establishing the hydrography of paleoseas; in particular, the reconstruction of the temperature, salinity and gas and hydrodynamics regimes is covered in great detail. However, the procedures for studying the character of the ancient dry-land regions are given insufficient treatment, and practically nothing is said about the action of paleorivers and ways of recognizing them. Zhizhchenko's large monograph contains sections dealing with the conditions of accumulation of organic matter in sediments and with the detection of the source beds. Unfortunately, however, the importance of paleogeographic studies for petroleum geology is not emphasized, and the practical recommendations for forecasting the occurrence of oil and gas are treated very cursorily.

Zhizhchenko can be credited with the development of several techniques of paleogeographic reconstruction, based on the synthesis of all available knowledge on the character and environment of deposition of marine formations of a certain age. We cannot but agree with him that paleogeography is a wholly independent branch of science calling for specialized training.

The work by Nalivkin, Vereshchagin et al. (1962), dealing with symbols and methodological recommendations for the Lithopaleogeographic Atlas of the U.S.S.R., offers advice on the preparation of paleogeographic documentation.

The Fifth All-Union Lithological Conference, which was held in Novosibirsk in 1961, discussed a special problem pertaining to the methodics of compiling various lithofacies and paleogeographic maps. The proceedings of the conference (published in 1963) include many articles dealing with this problem. A collection of papers presented at a session of experts of the U.S.S.R. Geological Council, with participation of the editorial board of the Lithopaleogeographic Atlas, was published in 1964. The session dealt with methods of paleogeographic investigation. The papers shed light on certain aspects of reconstructing geochemical and hydrodynamic conditions in ancient marine basins, treat some problems encountered in the study of paleorelief and paleoclimate, and discuss the experience gained from the application of biotic, ecological, paleomagnetic and other data to paleogeographic reconstructions.

Unfortunately, in recent years there have been very few publications on methods of paleogeographic investigation. While referring the reader to specialized works dealing with lithofacies maps, we feel obliged to comment that there are quite a number of other criteria, aside from facies and thickness, which complement one another and provide a clue to certain characteristics of past physiographic environments. It should also be borne in mind that however convincing an individual criterion might be, it is always preferable to consider it in conjunction with other criteria. Only by this means is it possible to ensure that the resulting information is highly reliable.

We shall also mention other studies and observations necessary for compiling comprehensive paleogeographic documentation.

### Paleoecologic Observations

Fossil remains of fauna and flora may be indicative not only of the age of
the rocks containing them, but also of the conditions of sedimentation.  Pa-
leoecology studies the habitat, way of life and other features of ancient organisms.  The methods of paleoecologic research are discussed by Gekker
(1957), Ivanova and Khvorova (1955), Maksimova (1955), Osipov (1955), Bel'-
skaya (1958), Zhizhchenko (1959), and others.

The mentioned authors present concrete examples of paleoecologic studies
of the habitats of ancient animals and plants in various basins.  Paleoecologic data are especially meaningful in reconstructing the paleogeographic environment if a facies analysis is performed concurrently.  It is very important to obtain information on the degree of preservation and conditions
of burial of organic remains, as this enables one to determine whether they
were redeposited or buried in situ.

The orientation and sorting of floral and faunal remains may be indicative
of the directions of currents and wave motions.  A change in the general
character of organic remains along the strike and in cross section suggests
a change in the paleogeographic conditions of sedimentation.

The history of the change in the earth's physiography cannot be studied
separately from the history of organic life.  This statement is borne out
by the original research of G. U. Lindberg, a well-known Soviet ichthyolo-
gist.  In particular, his recent monograph* tackles aspects of biogeography,
paleogeography and Quaternary geology in the light of the concept of the
common historical evolution of the fauna, flora, and the water bodies and
territories occupied by them.

Lindberg advanced and later substantiated the hypothesis that several
ancient river systems existed in the remote past.  He did this after studying the present-day distribution of ichthyofauna.  To prove his hypothesis,
Lindberg relied not only on data on the distribution of certain fish species,
but also on the history of evolution of the territory on which integral river
systems were situated.  He developed a technique of biogeographic analysis
of recent geologic events, proceeding from the concept that the organisms
and their environments are an integral whole.  The distribution and composition of organisms, which are representative of the process of shaping
the earth's surface, may serve as a criterion for verifying this scientific
approach.

Lindberg emphasizes that this method imposes stringent demands on the
quality of biogeographic indicators.  Not every group of organisms or territory meets these requirements.  It is essential that the group of organisms
selected for study be restricted to a specific habitat and that the territory
have geomorphological boundaries which served as an impassable barrier
to the group of organisms in question.  In studies of Quaternary history the

---

* G.U. Lindberg.  Krupnye kolebaniya urovnya okeana v chetvertichnyi period (Large fluctuations in ocean
  level during the Quaternary Period). — Moscow, "Nauka," 1972,  548 pages.

biogeographic indicators meeting these demands are typically fresh-water fish and river systems geomorphologically well isolated from one another.

Readers interested in a detailed description of the biogeographic method are referred to Lindberg's original works. We only wish to mention that he managed, by establishing the completely identical composition of the typically fresh-water fish fauna in rivers which are now isolated from one another, to show that certain present-day insular and continental rivers of the Far East, Western Europe and other regions of the world belonged formerly to a common integral system. Lindberg's hypothesis was corroborated by an analysis of the marine bottom relief, which revealed traces of inundated valleys which had once constituted a single river system.

A biogeographic study of the ichthyofauna of the rivers on the southeastern slopes of the Japanese islands and the southwestern slope of the Korean Peninsula and a comparison of this ichthyofauna with that of the Hwang Ho River, in conjunction with a geomorphologic analysis of the bottom relief of the Yellow and East China seas, made it possible to detect the inundated valley of the ancient Hwang Ho and the river valleys on the southwestern slope of the Korean Peninsula, which had formerly constituted integral parts of the drainage basin of this ancient river.

An important implication stemming from the tracing of the ancient counterparts of many present-day river systems is that the level of the World Ocean was repeatedly lowered to 200—300 m below its present level. This resulted in the temporary retreat of water from the continental shelves of all the continents and oceanic islands during the Quaternary. The major rivers thus cut their courses to the points of debouchure into the ocean across territories that had emerged from the sea. This is also indicated by submarine valleys of many rivers extending to the depths of emergence of the continental shelf, as well as by phases of low level of the ocean.

The presence of several well-delineated ancient inundated valleys of the largest Eurasian and North American rivers on the present-day Arctic shelf has been established. In particular, the Ob delta was located at the latitude of the northern extremity of Novaya Zemlya during the latest regressive phase of sedimentation.

We shall not go into the causes of these large geohydrocratic fluctuations in the level of the ocean. Lindberg discusses them for the Quaternary Period. We can only agree with his statement that there is every reason to believe that similar situations also arose during earlier phases of the development of the earth's crust.

Consequently, biogeographic studies, in combination with other data, make it possible to detect large fluctuations in the level of the ocean, and to gain knowledge of the characteristic features of the formation of the mouths of paleorivers, coastal plains and the offshore marine shelves, as well as the evolution of the organic life which once flourished there. All this undoubtedly links up with paleogeographic reconstruction, as well as with the regional and local accumulation of oil and gas.

Terrigenous Components in Paleogeography

The principles of reconstructing physiographic environments on the basis of terrigenous components are described in the works of Baturin (1937, 1947),

Pustovalov (1947), Rukhin (1940, 1961), Sarkisyan (1960, 1968), Kazarinov (1969), and others. These works deal with the composition, mode of transport and distribution of terrigenous complexes, and also describe how they should be studied.

Baturin focuses his attention on the mineralogy and grain size of clastic rocks, reflecting the dynamics of the hydrosphere and atmosphere, as well as on the distribution of the source areas supplying the detrital material. Paleogeographic reconstructions based on studies of terrigenous rock components take into account primarily the distribution of petrographic types in regions of denudation and deposition. Baturin recognizes terrigenous-mineralogic provinces — regions of sedimentation characterized by specific complexes of light and heavy minerals transported from one or several source provinces. Mineralogic analyses of the geological past, based on terrigenous components, may be used in petroleum geology not only for distinguishing barren strata, but may also serve as a method of paleogeographic reconstruction.

### Geochemical Data

Reconstruction of the physicochemical conditions of sedimentary environments and of the hydrochemical environments of ancient water bodies is based on geochemical criteria, in particular, on the composition of authigenic minerals in sedimentary rocks. Geochemical studies help to reveal the character and hydrochemical regime of individual regions of marine paleo-basins. The methods and procedures employed and results obtained from reconstructing the geochemical environments of sedimentation and secondary alteration of rocks, as well as the geochemical relationships between the latter and the organic matter, are dealt with in detail in the works of Pusto-valova (1933), Teodorovich (1947), Strakhov (1954), Gulyaeva (1956), Rodio-nova (1957), Uspenskii (1969) and many other investigators. We shall not go into a detailed discussion of the geochemical criteria used in paleogeo-graphic analysis but refer the reader to above-mentioned specialized sources.

### Structural Features of Rocks

Of great significance for paleogeographic reconstructions are the structural features of rocks, which provide an indication of the dynamic conditions of sedimentation. Bedding and lamination are the most common structures known to be due to the sedimentary environment.

Bedding is produced by alternation of sedimentary strata and layers of relatively homogeneous structure and composition; lamination refers to the internal structure of layers which are made up of individual laminae. The latter are usually a few millimeters or centimeters thick. According to Vassoevich (1948) and Botvinkina (1965), the lamina is the smallest recognizable unit layer in a sediment or sedimentary rock (the lowest unit of bedding). The distinctive features of layers and laminae are described by the above-mentioned authors.

Khabakov (1948) incorporated all indications of the dynamic conditions of sedimentation into the concept of dynamic paleogeography, which also

includes types of bedding. It is quite evident that structural features alone
are insufficient evidence for drawing conclusions about the genetic type of
a deposit. The genetic type may be established if other sedimentary characteristics are also taken into account. Nevertheless, bedding and lamination,
especially in the case of barren strata, may serve as major criteria for the
genetic classification of sedimentary formations, as well as in facies analysis and paleogeographic reconstruction.

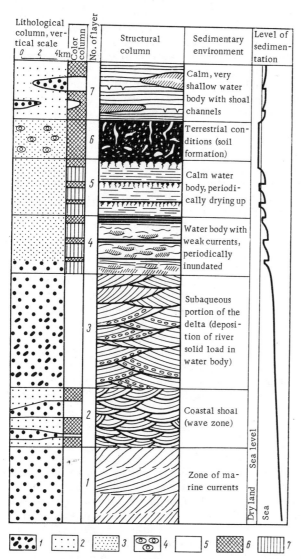

FIGURE 33. Structural column of an Upper Permian section, Ural region (Botvinkina; 1965):

1 — medium-grained sandstone with pebbles; 2 — fine-grained sandstone; 3 — siltstone; 4 — concretions;
5 — gray rocks; 6 — reddish-gray rocks; 7 — red rocks.

The character of the bedding may be indicative not only of the origin of the sediments but also of the direction of flow of the stream or wind which transported them.  In particular, it is sometimes possible to determine the direction of bottom currents in ancient basins from the bedding.

The significance of structural criteria in studying sedimentary environments was vividly illustrated by Botvinkina (1962–1965).  Since the structure may help to establish the genetic type of a deposit, it is worthwhile constructing a structural column (Figure 33) when studying core samples from boreholes and preparing lithologic columns.  This will make it possible to gain a better insight into the facies, and, consequently, the physiographic environments of sedimentation.  The sand layers 1 and 3 (Figure 33) display coarse cross-bedding of almost the same character.  However, closer inspection of the arrangement and structure of the cross-beds reveals that in layer 1 they were formed by marine bottom currents, whereas in layer 3 they are typical of the subaqueous portion of a delta.  Layers 2 and 7 are represented in the lithologic column by siltstones of very similar composition containing lenses of fine-grained sandstones.  However, structural features suggest that they are of different origin.

In the given example a thorough study of the structural features of the rocks in the section made it possible to specify the genetic type of each layer and to determine the trend of changes in the paleogeographic environment with time.  It can be seen from this example how misleading a formal approach to the study of the structural features of a deposit may be.  More specifically, if attention in this case had been focused only on the fact that the lower portion of the section is cross-bedded, while the upper portion shows horizontal bedding, this would have led to the erroneous conclusion that the sediments at the top were deposited in a deeper-water environment than those at the bottom.  It was only after a detailed examination of the structure of each layer that the reverse relationship and change of facies could be established.  The structural column shows graphically that sedimentation took place during a regressive phase, when a marine environment was successively superseded by nearshore marine, deltaic, shoal (first channel-type, then stagnant-water) and, finally, by terrestrial conditions.

The study of morphological and facies-genetic types of bedding and their classification are dealt with in great detail in the works of Botvinkina (1962, 1965), Vassoevich (1948–1958), Grossgeim (1950–1965), Rukhina (1953) and others.

### Climatic Criteria

In reconstructing the physiographic conditions of sedimentation during a certain period of the geological past one should not ignore the climate, which is an important component of paleogeography.  Data on ancient animal and plant communities and on the composition of the sediments are the main source of information about the paleoclimate.  Studies of the remanent magnetization of rocks reveal the migration of the earth's magnetic poles, which brought about changes in climatic zoning at the surface.

Strakhov (1945) assigns to climate the decisive role in determining the lithologic type and geochemical nature of a sediment.  The deeper we go into the details of the sedimentation process, the more evident is the intimate relationship between sediments and their physiographic environments of deposition. Various criteria are used to determine the geochemical features of rocks which originated under different paleogeographic conditions.

For instance, in a study of the geochemistry of the Upper Paleozoic humid zones of the U.S.S.R. (Strakhov, Zalmanzon and Glagoleva, 1959), the following criteria were employed:

1) the distribution of elements in the rock sequence: sandstone-siltstone-mudstone-marl-limestone. The distribution of Fe, Mn, P, V, Cr, Cu, Ni, Co, Ba, Sr, Be, Ge, Pl, Zn, and $C_{organic}$ was studied. It was found that the behavior of these elements as a whole is very sensitive to the physiographic environment of sedimentation and may sometimes reveal details of this environment which cannot be detected by conventional methods of facies analysis;

2) the balance of the mineral forms of iron. This depends on the physiographic environment of deposition and may be used for an analysis of relations between the geochemistry of the rocks and the facies environments in which they were formed;

3) the chemical composition of the carbonate material dispersed in rocks, especially if these rocks are clastic and argillaceous. The composition of this material is determined mainly during the stage of diagenesis, is connected with the balance of authigenic forms of iron and clearly indicates the secondary redistribution and chemical differentiation of individual components;

4) the composition of carbonate concretions occurring in deposits of humid zones. In most cases these concretions are the result of the redistribution of the sedimentary material during diagenesis. Studies of these concretions aid one to reveal the laws governing the redistribution process and the details of the physicochemical environment.

Bucher (1968), in his search for other features indicative of the paleoclimate, found it useful to study organic remains which impart a dark coloring to terrestrial deposits. He believes that the ratio of the thickness of the dark layers to the overall thickness of the terrestrial deposit within a given stratigraphic section may be an important parameter. In our opinion, the presence of coal is an even more specific indication of the paleoclimate. Plant remains, which are abundant in coal-bearing deposits, and the petrographic composition of the coals may be indicative not only of the climate, but also of a more specific landscape of coal accumulation.

Isotope paleothermometry, described by D. P. Naidin and other authors, is yet another way of determining past climatic conditions.

The striking similarity in chemical composition and time of maximum formation of all fossil fuels suggests their common origin and their close connection with periods and zones of maximum development of life, which varied as the physiographic environment, including climate, changed. Unfortunately, this connection is not given due attention in geological practice.

Occurrence of Coal

A coal-bearing formation is a rather accurate indicator of both climatic zoning and continental conditions of sedimentation. Coal formations are often widespread in oil and gas basins, especially on ancient alluvial-deltaic plains which later became shallow epicontinental seas.

Coal accumulation starts with the formation of peat which develop as the result of the moldering and partial putrefaction of marsh vegetation under conditions of high humidity and limited access of air. This takes place in lakes, swamps, oxbows and other inland water bodies where the conditions were favorable for the development of plants and the large-scale accumulation of their remains.

The petrographic composition of the coal reflects the conditions under which it was formed. The presence of vitrinite, for instance, suggests fluctuations in the water level at the time of burial of the plant material. Fusinite is formed with the participation of atmospheric oxygen. Incompletely putrefied wood remains could be its source material. The presence of fusain or fusain-xylain components in the coal points to the relative dryness of the peat bog.

The degree of pyritization and the sulfur content of coals make it possible to classify the water bodies in which the coal accumulated. A high content of sulfur and pyrite in the coal suggests that conditions in the water bodies favored microbiological reduction of sulfates; the hydrogen sulfide so evolved combined with iron to form pyrite. These phenomena point to the stagnant nature of the ancient water bodies, with predominance of reducing or strongly reducing environments in the benthic zone.

In flat areas of the paleorelief, in lakes and marshes, a single coal bed of considerable thickness was formed, whereas in river valleys and intradeltas the coal measures are multilayered but each coal seam is rather thin. Coal from floodplain marshes and oxbows contains some clayey material and is characterized by micro-intercalations of the individual components, this being due to fluctuations in the level of the water covering the peat.

Consequently, the character and structure of a coal-bearing formation are indicative of the physiographic environment during the period of coal accumulation. Unfortunately, in petroleum exploration and paleogeographic reconstruction very little attention is paid to the occurrence of coal, to the characteristics of discovered coal beds and to coal-bearing formations in general, notwithstanding the fact that they do provide an indication of the environment of deposition.


*Methods of Paleogeographic Reconstruction*

Identification of Alluvial Deposits

Most large oil deposits and pools of the lithologic and stratigraphic types are associated with alluvial-deltaic deposits. For this reason we shall discuss in detail methods of reconstructing buried river valleys and their mouths. A special branch of science — paleopotamology (from the Greek "potamos" — river) — is devoted to the study of ancient fluviatile deposits. Examples of such studies are the works of Goretskii (1964—1970) covering potamologic investigations of the paleo-Volga, paleo-Dnieper, paleo-Don and other ancient river valleys. The mentioned works examine the alluvial history of the major rivers of the Russian Plain and the stages of formation of their valleys on the basis of studies of alluvial formations and their relationship to moraines and fluvioglacial and marine deposits, and also information on tectonic and glaciotectonic movements.

Goretskii characterizes alluvial formations according to their bedding, thickness, lithofacies, structure, grain size, the position of river downcutting in profile, etc. The age of alluvial deposits is established with the aid of palynologic, paleocarpologic, conchologic, diatom and other analyses. The techniques used in these studies are described in special publications.

Alluvial deposits commonly contain sand strata with good reservoir properties. It is therefore of considerable practical importance to study these deposits.

A basic feature of alluvium is its association with erosion valleys. Rivers, particularly at the initial stage of a regressive cycle of sedimentation, cut their courses most readily in lowlying relief forms which are inherited or newly created by the geostructural pattern. Fluviatile deposits show both horizontal and vertical variations. They consist of various lithologic types. In view of the frequent alternation of facies, determined by climatic and orographic zonation and other factors, it is difficult to study such deposits.

Alluvium does, however, exhibit general distribution patterns which should be taken into account. These regularities have been investigated for recent and ancient alluvium in the well-developed valleys of rivers flowing in plains, and described by many investigators (E. V. Shantser, N. I. Nikolaev, G. I. Goretskii, Yu. A. Zhemchuzhnikov, L. N. Botvinkina, V. S. Yablokov, P. P. Timofeev and others). It is quite evident that the nature of alluvial deposits differs from river to river, being dependent on geotectonic, orographic, climatic and other factors. However, the structures of alluvial complexes have many features in common because of the recurrence of the same hydrodynamic regime of the flowing waters.

Studies of buried alluvium are based on the principle of uniformitarianism which is closely related to the method of comparative lithology. Factual data for studying the features of ancient alluvial strata should be assembled and generalized according to the following basic points:

classification of alluvial deposits according to their facies and other factors;

identification of individual facies (fluviatile, floodplain, oxbow) and determination of their interrelations;

analysis of the thickness of alluvial formations in river valleys, mouths and deltas;

detection of basal horizons of individual alluvial complexes;

determination and tracing of buried fluviatile deposits in section and in plan.

The formation of alluvium — a product of the transport and deposition of terrigenous material by rivers and streams — accompanies the formation of the erosion valley. This is why the facies and thickness of the alluvial sediments are closely connected with the formation of the valley itself. Alluvium has the following major facies: fluviatile, floodplain, and oxbow.

The experience gained by a group of geologists from the Institute of Geology of the U.S.S.R. Academy of Sciences while studying the alluvium of the Middle Carboniferous coal-bearing suites of the Donbas may serve as an example of the identification of lithologic types of alluvial deposits and their facies.  Timofeev (1954) classifies six of the numerous lithogenetic types encountered in the southwestern part of the Donbas as alluvial facies (Table 4).

TABLE 4.  Lithologic types and facies of alluvial deposits

| Lithologic type | Depositional environment | | Facies |
|---|---|---|---|
| Mudstone, siltstone, fine-grained with fine horizontal bedding and plant remains, partly well preserved | Slowly flowing or stagnant waters | | Inner portions of the floodplain |
| Siltstone with horizontal interrupted wave-like and lenticular bedding, containing plant detritus | | | |
| Sandstone, fine-grained with oblique, thin, convergent bedding and intercalations of rocks with fine horizontal interrupted bedding | | Floodplain | Part of the floodplain close to the river bed |
| Sandstone, medium- and fine-grained, with graded bedding and coarse unidirectional convergent cross-bedding | Flowing waters | | Fluviatile |
| Sandstone, coarse- and medium-grained, inequigranular, with graded bedding and coarse unidirectional rectilinear cross-bedding | | | |
| Gritstone, sandstone, coarse-grained, inequigranular, unsorted | | | |

Timofeev classifies as fluviatile facies a single paragenetic sequence of deposits with the following characteristic features:

the presence of coarse-grained inequigranular rocks in the lower portion of the stratum;

decrease in grain size and increase in degree of sorting from the base to the top of the stratum,

coarse unidirectional rectilinear cross-bedding at the base and coarse unidirectional convergent cross-bedding at the top;

multi-stage arrangement of oblique sets;

graded bedding within each cross-bed;

regular distribution of plant remains, from coarser fragments in the lower portion to fine detritus in the upper portion;

all the beds of the facies are separated from the underlying sediments of various horizons by an erosion surface;

presence of intraformational erosion.

Floodplain facies in proximity to the river bed, which usually overlie fluviatile deposits, possess the following characteristics:

presence of fine-grained sandstones and coarse-grained siltstones;

material much better sorted than underlying deposits;

decrease in grain size toward the top of the stratum;

fine convergent cross-bedding, cross wave-like and horizontal wave-like bedding;

bedding readily discernible due to presence of plant detritus and clayey material;

paragenetic relationship between fluviatile and floodplain deposits.

Deposits of the inner parts of the floodplain, which usually occur in the upper portion of the river alluvium, possess the following major features;

predominance of finely elutriated siltstones and mudstones;

absence of homogeneous mudstones lacking admixture of silt;

characteristic fine horizontal or slightly lenticular bedding;

abundance of plant remains, including well-preserved remains or imprints;

presence of siderite concretions;

close paragenetic relationship with fluviatile sediments and with soils and subsoils.

Fluviatile deposits form the base of the floodplain and constitute the bottom horizon of the alluvial stratum. Floodplain sediments make up the top horizon. Deposits of oxbow lakes, which are relatively rare, occur as lenticular bodies at the level of the bottom horizon. The relationships of these lithofacies groups are subject to the general laws governing river erosion and deposition.

It is sometimes very difficult to distinguish between fluviatile, floodplain, and oxbow alluvium, each of which, according to Shantser (1951), consists of a complex paragenesis of various facies and lithologic types of sediments. Moreover, the buried sediments of many alluvial facies, generally their upper portions, are removed by erosion, while the basal horizon and, partly, formations of the river bed are usually preserved.

According to Goretskii, deposits of the basal horizon, representing erosion facies, are of major significance in the study of ancient alluvial strata. Sediments of erosion or of the basal horizon are readily recognized by the admixture of gravel and pebbles or by the presence of coarse sand fractions. Determination of the thickness, grain size, degree of rounding, structure and petrographic composition of the clastic material constituting the basal and fluviatile sediments may be helpful in reconstructing ancient river valleys.

It is not always possible to distinguish individual facies and establish their interrelationships. Therefore, it is good practice to single out paragenetic complexes of alluvial rocks associated with the development of erosion valleys. This may be achieved by comparative investigation of alluvial sections in conjunction with study of the morphology of valleys produced by shifting streams.

Detection of River Valleys

River valleys may occupy vast territories. In the lower reaches of the Amu Darya and Syr Darya rivers, alluvium covers an area of 250,000 km$^2$. The near-mouth portion of the Mississippi River valley has an average width of about 100 km over a distance of nearly 500 km. The lower reaches of the Hwang Ho and Yangtze rivers, together with their deltas, occupy an area exceeding 500,000 km$^2$. The Volga delta covers an area of more than

$20,000 \, \text{km}^2$.   The lower reaches of the valleys of paleorivers are the localities where the ancient alluvium is best preserved.

In order to detect a buried river valley, the investigator should primarily study maps showing the tectonic or, preferably, paleotectonic zoning of the territory in question.   The ancient structural pattern which controlled the distribution of the hydrographic network is of major significance.

**Isopach maps** of individual terrigenous complexes may in this case be used as indicators of the ancient relief and of the general trend of the river valleys.   Such maps are based on data from exposures and boreholes.   River valleys can be recognized and traced to best advantage on a regional scale. The lithostratigraphic interval selected for constructing isopach maps is generally that of a single large regressive cycle of sedimentation or a single producing formation.   Maps of suites and individual horizons are prepared. It goes without saying that isopach maps must be supplemented by lithofacies maps, and the interpretation of both must be done jointly.

In places where ancient inheritance of an ungraded relief is encountered, the tectonic factor must be taken into account.   Only in those cases where the thickness of the sedimentary fill in the depressions of the buried relief is much greater than the average thickness of the area and there is erosion at the base of the studied series, is it possible to suggest an erosional origin. A decrease in thickness of the underlying rocks may serve as a confirmation.

A measure of the relative amplitude of erosion downcutting may be obtained by subtracting the mean depth of the base of the alluvium from the corresponding value of the paleorelief elevations.   The effect of the erosion factor should be determined by comparing the relief of the basement surface with the tectonic relief of some marker-bed in the sedimentary cover.

If borehole data are insufficient for tracing a buried river valley filled with thick terrigenous deposits, it is worthwhile resorting to geophysical methods, in particular, reflection shooting.   The experience of Urupov et al. (1961) in mapping an area of thick Lower Carboniferous terrigenous strata in the Volga Region near Kuibyshev has shown that this method is well suited to solving the problem.   The following seismic data may be used to assess the structure of the terrigenous stratum;

changes in the arrival time of reflections from boundaries below the stratum in question, if the latter undergoes abrupt change in thickness;

possibility of separate recording of reflections from the top of, bottom of, and interfaces within the terrigenous series in zones of great thickness;

change in the interval between reflections arriving from boundaries confining the terrigenous series of variable thickness;

change in the average speed of propagation of seismic waves in a series containing terrigenous rocks with low velocities and carbonate-sulfate rocks with high velocities, if the ratio of the thicknesses of these rocks varies.

The last two criteria furnish the most complete and unambiguous data on zones of terrigenous formations within carbonate deposits.

In bands of thick terrigenous deposits confined to ancient river valleys, especially in their estuarine parts, the changes in thickness show a certain regularity.   A gradual increase in thickness is due mainly to an increase in the quantity of sand deposits in the section.

Such bands and the erosion valleys containing them are usually oriented perpendicular to the shoreline of the marine basin.   If these bands expand

and their thickness increases abruptly at the same time, attaining maximum values in the coastal zone, this is a direct indication of the presence of a paleoriver delta in the region.

**Criteria for detecting buried fluviatile deposits.** Of all alluvial deposits, fluviatile formations are the most promising oil and gas reservoirs. The fluviatile facies has much more distinct features than the floodplain and ox-bow facies. However, it is difficult to detect even fluviatile deposits on the basis of core samples from individual boreholes, since the entire combination of characteristic features is seldom encountered within a single section. It is, therefore, necessary to make the most of various direct and indirect criteria. Bedding and grain size are the main clues to fluviatile deposits.

B e d d i n g. Cross-bedding due to the specific and irregular process of accumulation of the alluvial beds is the most characteristic structural feature of fluviatile deposits. Its form is closely related to both the mechanical composition of the sediment and the variability of streamflow. This is why the cross-bedding produced by river action differs considerably from other similar types of oblique bedding. Its main genetic feature is that the bedding takes the form of a series of parallel cross-beds mainly having a unidirectional slope of 25—30°. Oblique series are formed when a river deposits sand bars on its bed. These bars grow in the down-stream direction, their upstream parts being partly washed out. The continually shifting bars overlap one another.

The distribution of detrital material in a section of fluviatile deposits generally depends on the grain size of the sediments. The lower portion of the section consists of coarser-grained material, sometimes with indistinct cross-bedding. Toward the top the material becomes finer, and the cross-bedding is more evident. If the stream maintains a constant course, the boundaries between the series are parallel. The thickness and extent of a series depend on the streamflow. The weaker the flow, the smaller the extent and thickness of the sets. Cross-bedding is sometimes made more distinct by coarser-grained fractions or by the detritus of plant remains.

In river shallows the bedding often becomes wave-cross or even crisscross. Gently dipping cross-bedded sets, resulting from the build-up of sediments parallel to the slope of the shoal, may also be encountered here.

Cross-bedding sets range in thickness from 10 to 70 cm, seldom reaching 1 m. In section, fluviatile series generally taper from bottom to top, this being a feature which distinguishes them from marine deposits.

Since the oblique bedding is unidirectional in fluviatile cross-bedded sets, the direction of their dip is a helpful indication of the direction of streamflow. The azimuths are plotted on rose diagrams showing the directions of cross-beds or on diagrams showing both azimuths and dips. Procedures for studying cross-bedding are to be found in special manuals.

G r a n u l o m e t r i c   c o m p o s i t i o n. Buried fluviatile sediments are most readily recognized when they are made up of coarse clastic material; it is more difficult to identify fine-grained fractions. For this reason a layer-by-layer study of their grain size is of great importance. Fluviatile deposits are inequigranular and show considerable variations in grain size. They are coarser than floodplain or oxbow facies. They also display graded bedding, with a gradual transition from coarser to finer fractions. Abrupt changes in grain size from coarse-grained sand to siltstone may take place both within a single bed and from one bed to another within each set.

A river channel is divided into shallow (bars) and deeper (reaches) parts. The coarsest material is deposited on the bars where the flow is always faster, and the finer fragments are carried to the deeper parts. The location of bars and reaches changes constantly, and as a result the particle-size distribution also changes, the finer fractions overlying the coarser ones.

The recognition of fluviatile deposits on the basis of structural and granulometric criteria must be supplemented by studies of the variation in roundness of the grains, their degree of sorting and mineral composition. These and other criteria furnish the most reliable results, provided they are used in combination.

**Detection of paleodelta regions.** River runoff reaches its maximum in the delta region where the greatest accumulation of terrigenous material takes place. A number of authors (Nalivkin, Baturin, Pustovalov, Shvetsov, Samoilov, and others) have dealt with the processes of delta formation and with the recognition of their morphological features. They describe numerous examples of buried deltaic deposits of different ages. Paleodeltas are encountered in the Cambrian and Lower Silurian series of the mountain structures in Central Asia, in the Paleozoic and Mesozoic-Cenozoic deposits of the East European Platform, in coal-bearing strata of the European Upper Paleozoic, as well as in many younger terrigenous rock complexes.

Baturin's studies of the distribution of mineral associations in the terrigenous oil- and gas-impregnated strata of the Apsheron Peninsula have shown that this series is of deltaic origin, and that the sediments were supplied from the East European Platform. According to Khain (1954), the upper part of the oil-bearing formation there is locally separated from the lower part by a hiatus during which a north-south valley, more than 20 km wide and 70—80 m deep, was incised across the entire peninsula. This valley is filled with sand, gritstone and, partly, gravel.

A characteristic feature of the region occupied by the intradelta is its dissection by numerous distributaries which were formed as the main stream split up into smaller channels. Deltaic regions are not always conical or fan-shaped in plan. There are deltas with fairly rectangular outlines. In contrast to paleoriver valleys, which are almost totally obliterated, ancient deltas are much better preserved.

The fact that our large rivers are frequently inherited from older ones can sometimes serve as a basis for tentative forecasts of the location of paleodeltas. The East European Platform serves as an example. Large river valleys on its territory have generally followed a similar course during prolonged regressive cycles of sedimentation from the Riphean up to the present day.

Distinctive geomorphological, structural and lithofacies features are the main criteria in the search for paleodeltas to be investigated with the aid of geophysical surveys and drilling.

Geomorphology. The simplest way to locate a buried deltaic deposit is to drill through a modern delta built by a large river. Of practical significance in the search for commercial deposits of oil and gas are sedimentary formations of the remote geological past.

To this end, it is necessary to reconstruct the geological history of a river basin and of its deltaic region by proceeding backward in time and determining what changes took place. There are numerous examples of deltas that continued to exist for a rather long period of time, despite changes in the areas they occupied.

For instance, Recent sediments of the Mississippi River, which debouches into the Gulf of Mexico, overlie deltaic deposits of Pliocene age which, in turn, overlie older alluvial formations dating back to the Cretaceous. Repeated changes in the location of the mouth of this river led to the formation of a vast ancient alluvial plain measuring about 80,000 km$^2$ in area.

Structural features. The deepest negative structural elements of the first order, which could most frequently be filled with marine waters, can be recognized on a general tectonic map or on a map of the basement surface. Relatively small shallow depressions often link up with subsided basins. Whereas the basins could be occupied by a sea, the depressions might be part of the dry land. In this case it is natural to assume that such depressions were suitable localities for the incision of a river network or the course of a single large river. It is highly probable that the zone of transition from a depression to a basin is also the site of a transition from the river to the sea, i. e., the deltaic region. The dependence of the hydrographic network on the structural pattern and relief can be exemplified by the location of the mouth of a Middle Visean paleoriver in the eastern part of the East European Platform.

On comparing a map of the surface of the crystalline basement with the isopachs of the Yasnaya Polyana terrigenous series, one notes that the depression structures include zones or elongated belts of increased thickness. Lithofacies analysis shows that alluvial deposits are almost invariably associated with such depressions. It is true that in this case the isopachs are based on data from a large number of boreholes. However, if one knows the structure of the basement from geomorphological and geophysical data, one can outline a general pattern of distribution of the major deltaic zones.

An optimum forecast should be based on a combination of all geological data and on regions which have not undergone large-scale restructuring.

Lithofacies. In order to determine the lithofacies, one must rely primarily on available drilling data. In cases when such data are too scanty, geophysical surveys may play a significant role. Since buried alluvial deposits of increased thickness generally occur among carbonate rocks, they can be identified and traced by means of a seismic survey.

The Lower Carboniferous series in the Transvolga region near Kuibyshev was mapped by A. K. Urupov, Yu. A. Byakov and S. A. Shikhov. They reported that it was possible to assess the structure of the terrigenous series from the results of reflection shooting.

To date, geophysical methods of prospecting for oil and gas have been used mainly to solve structural problems. However, geophysical studies could also be useful in determining the lithology of extensive areas; this would undoubtedly speed up the assessment of the lithofacies characteristics of a given region. Petroleum geologists in conjunction with geophysicists should introduce geophysical exploration methods in order to detect and study large alluvial plains and buried paleodeltas.

Widespread among deltaic deposits are sand and silt channel fillings that are often good reservoirs for oil and gas.

Channel or shoestring oil pools were discovered in the Maikop oil-bearing region and described by I. M. Gubkin. Here the oil accumulated in sand lenses in the bed of an Oligocene river over a stretch of 10 km. The pool is made up of an alternation of clay and fine-grained sand with intercalations of coarse-grained sand and gravel. The thickness of sand lenses ranges from 0 to 50—60 m. The entire belt expands downstream.

Similar shoestring pools were later discovered in the U. S. A. (Kansas and Oklahoma), where lenses of Pennsylvanian (Lower Carboniferous) sands are 15—45 m thick, 5—10 km long, and 0.8—5.2 km wide. The sand lenses form bands sometimes more than 80 km long.

In the central part of the Greater Kinel arch of the Volga-Ural oil- and gas-bearing region there are meridional strips of sandstone within the Bob-rikov horizon. These strips are about 1 km wide. The sandstones in them account for up to 70% of the terrigenous series and are often coarse-grained. Their porosity is 30% and their permeability reaches 3 darcies. These channel sands contain structural-lithologic oil pools; production from wells often exceeds 100 tons per day.

In the same buried deltaic zone of the Middle Visean river, where the mentioned sand zones are widespread, a shoestring oil pool was discovered in fine-grained well-sorted quartzose sandstone in the Pokrovsk field. The latter occurs among clays and forms a narrow band, 250—1300 m wide, stretching for 10 km. The thickness of the sandstone is 5—10 m, and it pinches out toward the margins of the strip; its average porosity exceeds 20% and its permeability is 0.75—0.80 darcies. This pool is a typical forma-tion of a small stream.

Prospecting for shoestring pools is difficult in view of their relatively small size and the absence of surface shows. However, since they generally occur in deltaic deposits, it is necessary to detect paleodeltas as the first step. In structure the sandstones of shoestring pools differ little from typical fluviatile formations.

Deltaic formations begin to acquire a specific character on transition to the subaqueous zone of accumulation. In particular, there are intercalations showing indistinct oscillation ripple marks alongside cross-bedding. Ac-cording to Botvinkina, the various portions of the delta are characterized not so much by the bedding as by the stratification and the relations between individual elements of the entire deltaic complex.

The foreset beds of the prodelta form very gently inclined layers and show an internal laminated structure due to the action of water, both river and marine.

The bedding in the prodelta is generally criss-cross, the layers dipping in opposite directions on account of the direction of the diverging river distributaries as they debouch into a water body.

Since paleogeographic studies make it possible to conduct a purposeful search for large oil fields associated with paleodelta deposits, it is necessary to detect the regions where the latter occur. It goes without saying that in the search for buried deltaic formations as suitable natural reservoirs, one must make use of any information pointing directly or indirectly to possible estuarine zones of ancient river systems, in addition to tectonic, geomorpho-logic, lithofacies and other indications.

Present-Day Rivers as Indicators of Ancient Hydrographic Networks

A casual glance at a modern physiographic map of both hemispheres enables one to discern the relatively widely spaced network of major rivers of the world. The history of their development shows that they often have a character inherited from the main waterways of the remote geological past. Even after repeated alternations of transgressive and regressive phases of sedimentation, individual rivers continued to cut their courses in similar or sometimes the same directions and following the outlines which had existed before.

This has been convincingly proved by Soviet investigators who carried out extensive preliminary research prior to the construction of hydroelectric power stations on the Volga, Kama, Dnieper, Yenisei, Angara and other major rivers. The Gidroproekt geologists have established connections between these waterways and the rivers of Anthropozoic and Paleogene times. Their earlier history could not be studied because of the insufficient depths of the Gidroproekt boreholes. However, data forthcoming from deep drilling for oil and gas suggest that the record of some major rivers starts in the Paleozoic.

For instance, most of the major Early Carboniferous rivers, which flowed across the East European Platform, and whose mouths contain rich oil areas, coincide in general features with the valleys of present-day rivers (Volga, Kama, Belaya, Don, Dnieper, Pripyat). Buried deltas and prodeltas within the Jurassic and Cretaceous deposits of West Siberia, which also show the highest oil and gas contents, are usually located along the courses of present-day rivers (Ob, Irtysh, Taz, Pur, Chulym, Ket', etc.).

Carboniferous paleodeltas discovered in the U.S.A. and containing numerous oil and gas fields are either associated with or occur in the proximity of the deltas of present-day rivers (Missouri, Arkansas, Mississippi). Deposits of the Cenozoic deltas of the Mississippi and Rio Grande, which nearly coincide with the present-day deltas of these rivers, are known for their exceptionally high productivity. A similar situation exists at the mouths of the Orinoco in Venezuela, the Irrawaddy in Burma, the Niger in Nigeria, the Ganges and Brahmaputra in Bangladesh, the Nile in Egypt and some of the rivers in Indonesia, all of which had experienced steady subsidence.

All this is by no means accidental, but perfectly regular. The life of rivers is directly connected with the tectonic evolution of the earth's crust, and they always flow toward large depression zones. The latter include primarily zones of transition from platforms to geosynclines, which include regions of subsidence on the margins of platforms (Khain, 1970). A major role is played by basins on the periphery of crators, which have subsided during entire tectonic cycles or even several consecutive cycles, i.e., for millions of years (the Volga-Ural region, depressions on the southern and western margins of the North American Platform, etc.). A second type of depression zones is found at the transition from continental platforms to young oceans (coasts of West Africa, South America, southern Australia, western Hindustan, etc.). A third type includes large downwarps of the basement (Caspian Plain, Gulf of Mexico, etc.). A fourth type is represented

by the depressions of inland and marginal seas located in young Alpine oro-
genic zones (Black Sea, Southern Caspian, Caribbean, etc.).

   The above types of zones, which Khain lists among the major belts of oil
and gas accumulation, are regions of intensive and prolonged subsidence,
which have not experienced inversion.  They were or still are filled by seas.
It is, therefore, natural that many rivers today flowing into these depression
zones have inherited the features of their precursors.  Therefore, if one
takes into account the shift of ancient shorelines of marine basins, the location of a major present-day river network may serve as a reference in
the search for "fossil" river mouths.  In order to obtain a more complete
assessment of the productivity of a certain paleodelta, it is necessary to reconstruct the history of its development both in space and time.


## 10.  PALEODELTAS AS MAJOR RESERVOIRS
     OF OIL AND GAS

   Soviet geologists were the first to note the association of oil-rich areas
with deltaic formations.  Baturin showed in 1937 that the productivity of the
famous Apsheron region is related to Miocene-Pliocene deltaic deposits of
the paleo-Volga.  In 1955 the present author detected the mouth of a paleoriver in the Visean stage (Lower Carboniferous) in the Transvolga Region near
Kuibyshev, and later several Paleozoic deltas containing major accumulations
of petroleum were discovered in the Ural-Volga region.

   In recent years mouths of ancient rivers of Jurassic and Cretaceous age
were discovered in West Siberia and on the Mangyshlak Peninsula, where
large hydrocarbon deposits are located.  We shall discusss some of them
later in this book.  We shall only remark in passing that, since the publication of Baturin's work 35 years ago, paleodeltas have not been given due
attention by petroleum geologists.  No scientific or industrial enterprise
has yet assessed the possibilities offered by the detection of river-mouth
zones in oil and gas basins for forecasting and for increasing the effective-
ness of exploration work.

   These opportunities are being used to advantage on an ever increasing
scale in the U.S.A.  It is true that American geologists have adopted an
empirical rather than a scientific approach to assessing the productivity
of ancient deltas.  However, they have met with success and their experience
is instructive.

   It is especially difficult to reconstruct paleogeographic environments in
deltaic regions, where intensive sedimentation continued over tens of thousands or even hundreds of thousands of years on vast territories, while the
hydrodynamic regime of the river and the position of the shoreline changed.
The accuracy of reconstruction depends on the extent to which the territory
in question has been covered by drilling.  However, even in cases where
large numbers of wells have been drilled, the paleogeographer is often obliged
to rely on logging data.

   Fischer, Saitta and Phares (1971) studied the oil-bearing Bartlesville
sandstone (Pennsylvanian) in eastern Oklahoma.  They made original use of

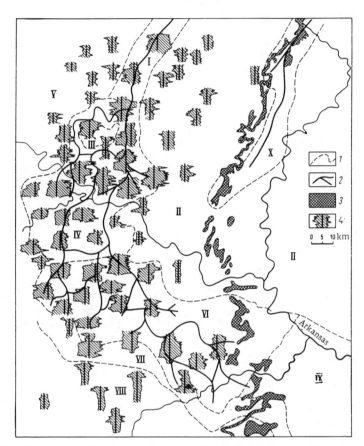

FIGURE 34.  E-log patterns and environmental reconstruction of deltaic elements of Pennsylvanian age in eastern Oklahoma (G. Fischer et al., 1971):

1 − environmental boundary;  2 − distributary pattern;  3 − Bluejacket outcrop;  4 − log location.  Deltaic elements:  I − alluvial valley;  II − marginal marine bay − basin;  III − bay;  IV − upper delta plain; V − marginal shelf;  VI − lower delta plain;  VII − lagoon;  VIII − marginal marine depositional plain; IX − distributary mouth bar;  X − minor deltaic area.

electric-logs for paleogeographic zoning of this deltaic region. They established its ancient geographic elements according to the shape of the E-logs which indicate the environmental sequences of the Bartlesville horizon in different parts of the delta (Figure 34). The data of exploration geophysics were correlated with the findings of textural, granulometric, petrographic and other studies of core samples. A combined study of the thickness, composition and texture of deposits in individual parts of the delta made it possible to distinguish the alluvial valley, upper delta plain, lower delta plain, marginal shelf or submerged sheet sand, marginal marine basin and marginal marine depositional plain. Each of the above elements is characterized by a particular E-log pattern, is associated with a specific paleogeographic element, and is the result of a particular depositional process.

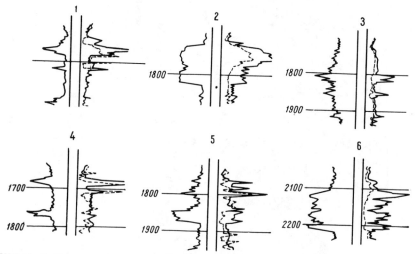

FIGURE 35. Fundamental types of SP log patterns. Individual environmental sequences are illustrated (Fischer et al., 1971):

1 — transgressive bar; 2 — major distributary; 3 — crevasse splay-bay; 4 — regressive sheet sand — transgressive bar; 5 — prodelta sand — transgressive bar; 6 — natural levee and minor channel sands.

Figure 35 shows fundamental types of SP log patterns for various groups of beds. Interpretation of the logging data correlates well with changes in the composition of the sandstone. The map showing the distribution of the Bartlesville sandstone (Figure 36), prepared by the above-mentioned authors on the basis of thickness measurements from over 5000 wells, also reflects the character of the braided paleostream network. The direction of the latter was determined from the linear orientation of the sand bodies, cross-bedded sands, fragments of wood, traces of erosion and other features. The paleogeographic reconstruction was corroborated by studies of natural outcrops of analogous sandstones east of the territory under study. The same isopach map, together with other data, shows that the region in question was a shallow-water shelf with coastal marshes, lagoons and bays. A large plain river passed through this territory.

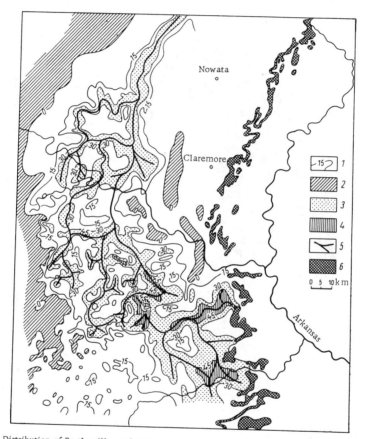

FIGURE 36.  Distribution of Bartlesville sandstone.  Map shows suggested distributary pattern and paleocurrent directions from outcrop studies (Fischer et al., 1961):

1 – isopach contours in m;  2 – zero thickness area;  3 – 30-m thickness area;  4 – 45–60-m thickness area;
5 – direction of paleocurrent;  6 – Bluejacket outcrop.

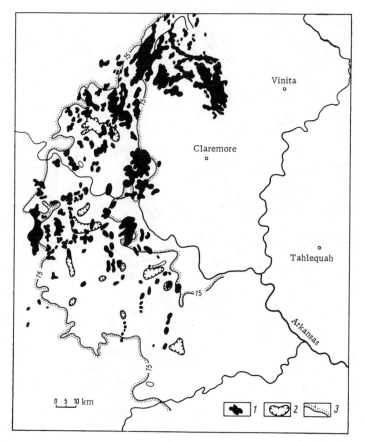

FIGURE 37.  Distribution of oil fields within Bartlesville sandstone in eastern Oklahoma.  Pattern shows concentration of oil on updip pinch-out of sandstone (Fischer et al., 1971):

1 — oil field;  2 — depression contour;  3 — area where sandstone thickness exceeds 15 m.

In the northern part of eastern Oklahoma there are deposits of two major waterways which split up into distributaries and diverging channels further to the south and east.  The basal channel sandstone throughout the territory of the Bartlesville paleodelta is underlain by marine argillaceous rocks; in some places erosion downcutting is observed.  The "marginal marine depositional plain" identified by Fischer is, in all probability, a beach where sands brought by the river and reworked by wave action were deposited. These sands occur in the south and west of the territory.  They are mainly sheet sands but they sometimes have a lenticular form (Figure 37).

More than 2500 oil deposits have been discovered within the Western Interior basin (U.S.A.), which includes Oklahoma.  Many of them are associated with zones of lithologic pinch-out, unconformity, truncation and confinement.

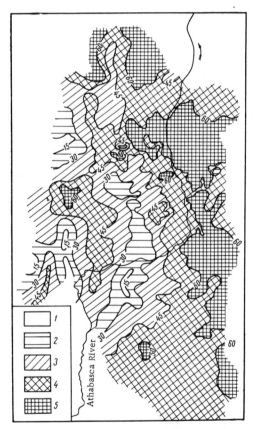

FIGURE 38.  Isopach map of the Athabasca tar sands, McMurray formation (Carrigy, 1971):

Areas with sand thickness: 1 — less than 15 m; 2 — 15 to 30 m; 3 — 30 to 45 m; 4 — 45 to 60 m; 5 — greater than 60 m.

American geologists have arrived at the interesting conclusion that, had they possessed information on the depositional environments and distribution patterns of the productive sandstones at the time of the first petroleum discoveries in eastern Oklahoma, there would have been no need to drill thousands of "dry" wells.

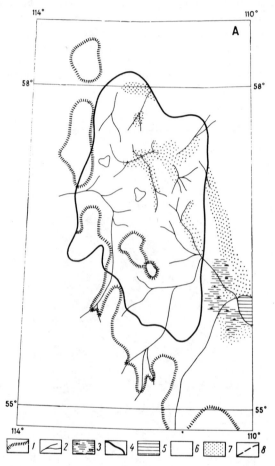

FIGURE 39.  Paleogeographic reconstruction of the depositional history of oil-impregnated Lower Cretaceous formations in northeastern Alberta (Canada).  Stage A: Lower McMurray (Carrigy, 1971):

1 — escarpment;  2 — stream;  3 — marsh;  4 — approximate boundary of tar sands;  5 — water;  6 — land;
7 — sand;  8 — approximate limits of McMurray sedimentary basin [not marked on figure in Russian original].

Another example of paleogeographic studies is Carrigy's (1971) generalization of material on the Athabaska oil field in Canada, where the McMurray tar sands (thickness 0—100 m) contain about 95 billion tons of heavy oil. The geometry of the sand bodies and the lithofacies composition of the rocks suggest that this formation was deposited within a "classical delta" in Early Cretaceous times.   Later it was rapidly buried beneath sediments of a trans-gressing sea.   The isopach map (Figure 38) clearly shows changes in thick-ness of the McMurray sands, which are mainly due to the hydrodynamics of streamflow in the subaerial and subaqueous portions of one or several del-tas, carrying and depositing the sands, rather than to the eroded surface of the underlying Devonian limestones.   The sands pinch out gradually in a northwesterly direction and are replaced by clays.

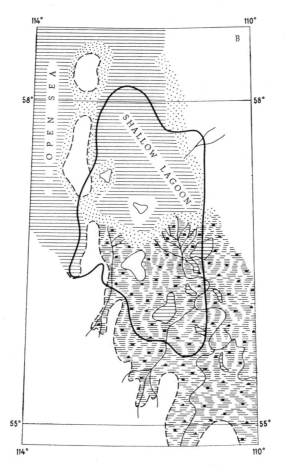

FIGURE 40.  Reconstruction of Stage B: Middle McMurray (Carrigy, 1971). (Legend as in Figure 39.)

The distribution of oil within the sands is controlled strictly by the pri-mary porosity and permeability.   Clean sands, most commonly of fluvial origin, contain maximum amounts of oil.   No oil is present in the poorly sorted argillaceous nearshore-marine and deltaic deposits.   The unique

Athabaska oil field provides convincing support for the assumption that paleoriver mouths which have undergone steady subsidence are exceptionally promising sites for oil accumulation.

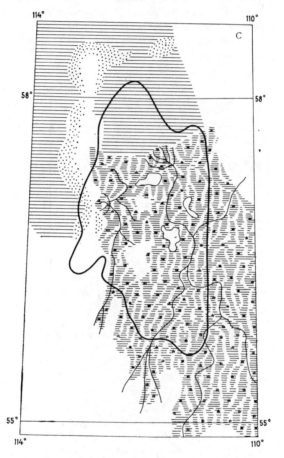

FIGURE 41.  Reconstruction of Stage C:  Upper McMurray (Carrigy, 1971). (Legend as in Figure 39.)

Carrigy's lithopaleogeographic reconstructions have made it possible to gain an insight into the geological history of the basin.  The subsidence of a vast territory between the Precambrian shield in the northeast and the Rocky Mountains in the southwest resulted in the formation during the Early Cretaceous of a lowland plain with a well-developed drainage system which carried coarse fluviatile material into numerous lakes (Figure 39).  Further subsidence caused expansion of the subaqueous basin.  The middle and upper members of the McMurray formation were deposited in alluvial and lacustrine-palustrine environments (Figures 40 and 41).  The intensified marine transgression from the northwest inundated the Athabaska region, and as a result the deltaic deposits were buried under the thick series of marine clays of the Clearwater formation (Figure 42), which formed the cap-rock. Table 5 lists the main characteristics of the producing sandstones in the mentioned deposit, indicating their genetic nature.

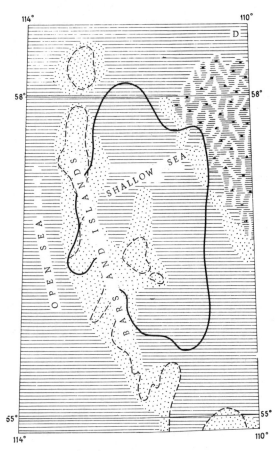

FIGURE 42. Reconstruction of Stage D: Lower Clearwater formation (Carrigy, 1971). (Legend as in Figure 39.)

Table 5 summarizes certain criteria which enable one to recognize an-cient deposits of zones transitional from alluvial-deltaic to prodelta. The example of the giant Athabaska oil accumulation is another confirmation of the connection existing between the formation of gigantic deposits and the sediments of ancient deltas.

TABLE 5. Features of environments of sandstone deposition in the Athabasca tar sands (after Carrigy, 1971)

| Parameter | Deposit | | | |
|---|---|---|---|---|
| | pre-deltaic | deltaic | | post-deltaic |
| | alluvial | lacustrine-lagoonal | | marine |
| | point bar and channel sands | foresets | subaqueous topsets | barrier bar |
| Texture | Medium- to coarse-grained well-sorted sand | Very fine- to fine-grained well-sorted sand | Silt and very fine-grained sand | Sand, silt, and clay, poorly sorted, locally clean and well-sorted |
| Sand composition | Quartz ≈95%, K-feldspar <5%, muscovite <1% | Quartz ≈90%, K-feldspar <5%, muscovite <5% | Quartz ≈90%, K-feldspar <5%, muscovite <5% | Quartz ≈50%, chert and volcanic rock fragments ≈25%, glauconite ≈20%, feldspar ≈5% |
| Clay composition | Kaolinite, illite | Kaolonite, illite | Kaolonite, illite, montmorillonite | Montmorillonite, illite, kaolinite, chlorite |
| Organic matter | Mummified logs, abundant comminuted carbon; some thin coal and lignite beds present | Abundant comminuted carbon | Macerated vegetable matter in clay matrix | |
| Mineral cements | Siderite (rare); pyrite nodules (rare) (up to 6 inches diam.) | Siderite (rare) | Siderite (common) and calcite | Siderite (common) |
| Fossil content | Spores and pollen | Spores and pollen | Some lenticular beds of brackish water gastropods and pelecypods; agglutinated foraminifera; pollen and spores; hystrichospherids | Calcareous foraminifera, radiolaria (pyritized), sponge spicules (pyritized), hystrichospherids, diatoms (pyritized) |
| Sedimentary structures | Medium scale (25 cm), straight, high-angle cross-beds (common) | Large-scale (30 m), low-angle (5-7°), foreset beds; small scale, micro-cross-lamination; burrows and castings (common) | Lenticular sand layers in silt; crab(?)burrows (rare) | Horizontal and vertical tubes filled with clean white sand |
| Bedding | Lenticular cosets of cross-stratified sets | Silt beds (1 cm thick) interbedded with very fine-grained sand beds (10 cm thick) | Laminated micaceous silt with lenticles of sand | Bedding destroyed by animal activity |
| Thickness | 2-7 m | 20-30 m | 5-10 m | 2-10 m |
| Location | Lower one third of the McMurray formation | Lower two thirds of the McMurray formation | Upper one third of the McMurray formation | Basal Clearwater formation |
| Oil content | Very good, uniform (up to 35% by volume) | Fair to good, uniform (20-30% by volume) | Poor (5-10% by volume) | Very poor, variable (local bodies of rich tar sand) |

The papers presented at the symposium on "Sedimentation in Deltas and Oil and Gas Accumulations" consider deltas as "closed systems" meeting the three basic requirements for the formation of commercial oil and gas accumulations: source, reservoir, and trap.

As R. Weimer aptly put it, ancient deltas may become "happy hunting grounds" where the geologist may successfully apply his skills and abilities. We may add that these "hunting grounds" contain unique giant fields where the "hunt" calls for detailed paleogeographic reconstructions. Carrigy presents an example of an analysis of the basic parameters which characterize the fluviatile deposits of the McMurray formation (Figure 43).

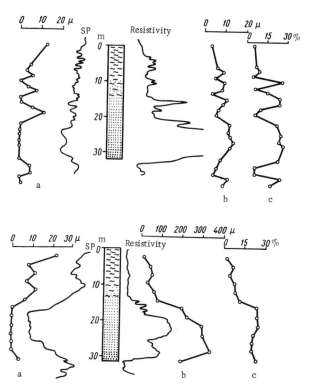

FIGURE 43. Basic parameters of two oil-saturated lacustrine-deltaic units of the McMurray formation in the Athabaska tar sands (Carrigy, 1971):

Curves: a – clay content (up to $2\mu$, % by weight); b – grain size (median diameter in $\mu$); c – oil content (vol. %).

Over the greater part of the Athabaska field the section of the producing sandstones is characterized by similar variations of the mentioned parameters. However, in the northeastern part the SP curve is a mirror image of the curve for the deltaic deposits. The fluviatile deposits there are replaced by marine ones.

Consequently, logging data, in combination with other criteria, may be used to detect fluviatile and marine sandstones.

*Paleogeographic Data and the Search for Lithologic Pools*

As pointed out above, lithologic traps account for nearly half of the total oil production in the U.S.A. In the U.S.S.R. the possibilities appear to be equal if not greater. However, in order to realize these possibilities it will be necessary to employ new and up-to-date exploration methods suited to the specific conditions of the given area.

To date, lithologic and stratigraphic pools in the Soviet Union have been explored only to a fairly limited extent. Therefore, these yet undiscovered fossil fuel deposits could well add to the proven reserves, especially in long-worked oil regions.

King (1971) identifies the following groups of lithologic and stratigraphic traps:

traps associated with variation in the properties of the reservoir and facies change to impermeable rocks;

traps in which the permeable reservoir is truncated by a stratigraphic unconformity and sealed by an overlying impermeable bed;

traps formed by transgressive and regressive permeable terrigenous deposits occurring above or below a stratigraphic unconformity;

bar and channel traps;

porous reefs and other lenticular carbonate bodies.

Exploration for these types of traps is much more complicated than that for structural traps. It can prove successful, but may call for considerable outlays on drilling operations. Less drilling is required if relatively detailed paleogeographic maps are available. The experience gained by American geologists shows that paleogeographic analysis is the principal method of prospecting for non-structural pools. The greater the detail of the analysis the more successful the exploration.

Levorsen (1970) distinguishes an independent type of combined traps, partly structural, partly lithologic. Any type of trap must obviously have rocks which are capable of accumulating hydrocarbons. But this is not enough. It is also essential that there should be source beds in the vicinity, whose formation is also controlled mainly by the physiographic and biogeographic environment of sedimentation.

There is not much available information on how one should go about exploring for oil and gas pools associated with non-structural traps. In this connection recent works by Maslov (1968), Gostintsev and Grossgeim (1969) and others merit mention. Obviously, it would be unfair to expect a straightforward and unambiguous answer from the investigators, especially since the conditions of formation and distribution of such traps are much more complex and diverse than those of structural traps.

Maslov (1968) provides the theoretical foundations and methodological principles of exploration for lithologic and stratigraphic pools in terrigenous formations. He discusses the genesis and morphology of traps of this type in relation to the facies conditions of sedimentation, subsequent tectonic deformation, or stratigraphic unconformities. He emphasizes particularly that the information necessary for justifying a search for such pools must be more comprehensive than that required for forecasting structural pools. A considerable part of the factual material presented in Maslov's book relates to exposed sections in the Northern Caucasus. Unfortunately, the

paleogeographer-petroleum geologist usually has to rely on well logs and
a limited number of core samples.

In 1969 a small collection of papers entitled "Paleogeographic Methods
of Research in Prospecting for Lithologic Oil and Gas Pools" was published
in issue 278 of Transactions of VNIGRI [All-Union Petroleum Scientific Research Institute of Geological Exploration].  These papers deal with the
principles of documentation, contributing to a better understanding of paleogeographic environments of sedimentation, rather than with the application
of paleogeographic studies.  Most of the papers in the collection discuss
techniques of studying sections on the basis of natural exposures rather than
drilling data which are in most cases the only source of information on oil-
and gas-bearing formations.

The mentioned collection (V. A. Grossgeim et al.) emphasizes that paleogeographic methods are of major importance for forecasting pinch-out zones
of reservoirs.  It is recommended that in planning exploration for lithologic
and stratigraphic pools one should use the same plots as employed in prospecting for anticlinal structures.  However, they must be supplemented by
maps showing sand content, currents, terrigenous-mineralogic provinces,
granulometry, roundness of sand grains, degree of sorting of sediments, etc.,
which make it possible to judge the provenance and routes of transport of the
terrigenous material along the bottom of the basin.

A special symposium of the Eighth World Petroleum Congress discussed
methods of exploring for lithologic and stratigraphic oil and gas traps.
R. E. King believes that until direct methods of oil exploration are developed
or geophysical methods are improved, the search for stratigraphic traps
that are not associated with reefs will be expensive due to the necessity of
wildcatting.  Reefs may be detected with the aid of seismic surveys or, less
successfully, using gravimetric data.

A large group of Soviet scientists (A. G. Aleksin, I. Kh. Abrikosov,
V. A. Grossgeim, I. P. Zhabrev, K. S. Maslov and others) who presented a
paper to the Congress are of the opinion that various types of lithologic traps
are actually formed during the process of sedimentation under the controlling influence of the physiographic environment.  Lithologic traps in marine
environments tend to be formed in zones with an active hydrodynamic regime
characteristic of shallow and nearshore parts of basins: beach strips, spits,
bars, reefs, etc.  Stratigraphic traps are formed after the deposition of the
sediments and result from tectonic movements causing gaps in sedimentation, erosion, and subsequent unconformable contacts between the reservoirs
and the overlying impermeable rocks.

When dealing with the methods of exploration for lithologic and stratigraphic traps, the authors of the paper point out that maps depicting the
paleogeographic and paleotectonic environments are of greatest interest at
the stage of regional studies.  In the exploration stage large-scale maps
are prepared, showing the structure of local traps associated with lithologic
pinch-outs and stratigraphic unconformities of the producing horizons; lithologic maps showing reservoirs, their thickness, etc. are also of great help.

One might have expected that such an assessment of the role and significance of physiographic conditions of sedimentation of producing formations
would be accompanied by concrete recommendations supported by examples

from practical experience.  However, the authors of the paper state in the conclusion that theoretical substantiation of the conditions of formation and spatial distribution of traps and oil and gas pools of this type, especially the methods of exploration, is still insufficiently developed.

This conclusion further confirms the need to introduce paleogeographic studies in exploration work as soon as possible.  It is high time to proceed from explanations and arguments in favor of the usefulness of paleogeographic reconstructions in petroleum geology to their purposeful application.  This method is particularly successful in locating zones of maximum oil and gas accumulation, in revealing distribution patterns of reservoir rocks and paths of migration of hydrocarbons, in pinpointing areas that should be covered by drilling, etc.  If one possesses reliable paleogeographic information one can more readily decide on an appropriate method of exploring for lithologic traps.  Naturally, this information must be used in combination with the results of other studies.

H. H. Christie (Canada), speaking at the mentioned symposium of the Eighth World Petroleum Congress, gave an example of a very rich litholog-ic-stratigraphic oil trap — Mitsue — in the northern part of Central Alberta.  Oil is produced here from the Middle  Devonian Hillwood sandstone.  The field is about 55 km long and up to 13 km wide.  The producing sandstone takes the form of lenses and alternating beds of fine- and medium-grained varieties which were deposited in the branching part of a bird-foot delta.  The average thickness of the oil-impregnated sandstones is 4 m, locally reaching 11 m.

The Hillwood horizon is an intricate combination of deltaic sandstone and bar sandstone.  The absence of positive structures in this region was formerly considered as almost conclusive proof of its unproductiveness.  However, a typical lithologic-stratigraphic pool was found in this area.  The SP-logs of the Hillwood sandstone match those of known fluviatile, deltaic and prodelta deposits.

Another example of a huge buried trap was given in the paper by G. A. Young (Venezuela).  In the East Venezuela basin, 150 km south of the Caribbean Sea, the Oficina area is located, and not far from the Orinoco River mouth, there is an area in which 97 oil fields were discovered, 20 of them with reserves of 8,000,000 m³ each.  Some 5560 wells have been drilled on a territory of approximately 19,000 km², and the current production exceeds 70,000 m³ per day.

The oil was discovered in sandstones occurring among siltstones, lignites and paralic and deltaic mudstones.  The producing sandstones have a decidedly lenticular structure.  Their thickness ranges from 0.6 to 12 m, locally reaching 60 m.  According to Young, it is difficult at present to ascertain whether they are the sediments of isolated or lunate bars, of natural levees or formations of a barrier beach.  The section contains more than 100 sand sheets or lenses which were definitely formed in the coastal zone.  The area of their occurrence is complicated by tectonic dislocations with amplitudes of 214 m or more.

Figure 44 presents the distribution pattern of typical sandstones in the Oficina area, while Figure 45 shows their cross section.  In Venezuela geologists use the following methods in the early stage of exploration:

geophysical prospecting, structural drilling, stratigraphic studies, and detailed subsurface mapping. The search for traps is a continuous process of investigation and reappraisal of data. The results of previous surveys, reassessed in the light of new findings and concepts, continue to serve as the basis for discovering new traps. It was found that the most effective method of exploration is the combination of reflection shooting and structural drilling. Most of the large tectonic dislocations and the majority of the oil fields were discovered by this method.

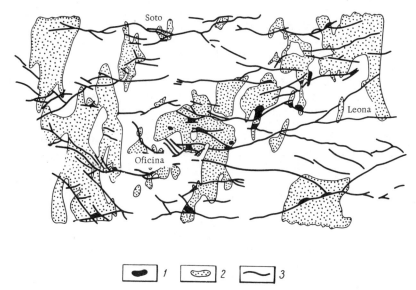

FIGURE 44. Distribution pattern of channel sands in the Oficina area (Young, 1971):
1 − oil pool; 2 − sand thickness exceeding 1.5 m; 3 − established faults.

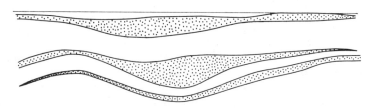

FIGURE 45. Cross section of the sands of the Oficina formation in the Ostra field (Young, 1971)

Two methods were employed to detect combined structural-stratigraphic traps: drilling of special wells in search for stratigraphic pools, and preparation of semiregional isopach maps. The first method involved drilling wells along large faults detected by seismic exploration in order to locate and assess additional sand lenses. In the second method semiregional isopach maps were prepared for each sand lens in order to determine the boundaries of each channel trap.  It was assumed that sites where channel sands are traversed by faults serve as traps with conditions favorable for oil accumulation.  After the detailed mapping many new traps and possible extensions of channel sands, covered by drilling, were plotted on the maps. Forty-five out of 100 traps were found to contain oil.

In view of the voluminous data available on the Oficina area, it was neces- sary to use computers to process them.  The computer, after being fed in- formation on the wells, can produce maps of any one of the hundred sand- stone beds (lenses) in the producing formation.  Despite the many difficul- ties encountered in the detection and tracing of meandering channel sands, the outlay involved may be justified, since the strike rate was, on the aver- age, 51%.  In 1970, 12 exploratory wells were drilled in channel sands, at some distance from the previously discovered fields; four of them produced oil.

The number of exploratory wells is another criterion of success.  More than 700 wells were drilled on the Oficina area during the 15 years of ex- ploration for stratigraphic pools.  Twelve major and numerous minor fields were discovered.  The strike rate ranged between 33 and 63% over the years.

Lithologic, stratigraphic and certain other types of traps may in fact be classified as nonstructural traps.  Halbouty (1972) believes that facies changes and unconformities lead to the formation of buried erosion or ag- gradation structures of the reef type, protruding sand bars, channel sands, and similar relief forms which may develop into nonstructural traps.  The mentioned author classes stratigraphic traps, paleogeomorphologic traps and traps of unconformity surfaces in this category.

Paleogeomorphologic traps are produced in areas where sharp subaqueous or subaerial features of the ancient relief are buried beneath younger de- posits of a different lithologic composition.  These include buried reefs, river channels and distributaries, rills formed by undercurrents, sand bars and elevated portions of erosion surfaces.  Traps of this type may occur both above and below the erosion surface.  They are directly or indirectly associated with buried relief forms.

Traps of unconformity surfaces are formed in areas where an imper- meable layer above an erosion surface is in contact with the underlying reservoir which must be sealed by impermeable rocks.  Such traps usually occur in combination with angular unconformities and result from the inter- play of processes of sedimentation, tectonic movements and erosion.  Their size ranges from small to giant, such as the well-known East Texas oil field.

Halbouty, speaking of the possible quantity of nonstructural traps in oil and gas basins, maintains that they are more numerous than structural traps, since they are formed under recurring conditions of sedimentation, usually preceding tectonic movements.  The main problem is how to detect them. He emphasizes that petroleum geologists must see to it that coordinated ex- ploration for nonstructural traps becomes a vital and integral part of explo- ration work.  They should resort to all methods of crustal studies which can shed light on ancient sedimentary environments favorable for the for- mation of nonstructural traps.  The significant role of paleogeographic studies in this context is quite evident.

*The Search for Lithologic Traps*

In order to select the ways and means of detecting lithologic pools, it is
first necessary to establish the zonality of oil and gas accumulation based
on detailed reconstruction of the paleogeographic environment in which the
producing formation was deposited.

It follows from the preceding sections that rather favorable conditions
for the formation of lithologic pools arise at river mouths. During the
period of sedimentation the rivers take an active part in sedimentogenesis;
in a buried state, their distributaries (especially those of prodeltas) are
convenient paths for the migration of fluids from the submerged portions
of sedimentary basins toward periphery. When such migrating fluids encounter lenses or other forms of sand formations, they fill them with oil
and gas. Such productive traps are primarily formed in nearshore, prodelta, deltaic, bar and alluvial-deltaic arenaceous rocks.

In poorly studied regions it is recommended that one approach the detection of paleoriver mouths by analyzing the general structural pattern of the
area. The ancient structural pattern, which was responsible for the distribution of the hydrographic network, may aid one to discern the main waterways and their directions. The structural factor is mainly expressed in the
distribution of sediment thicknesses. Therefore the trends of paleovalleys
may be revealed by isopach maps of individual terrigenous rock complexes.

Geostructural features indicating buried estuarine zones are recorded
in the following way: the most downwarped structural elements of the first
order, which are generally occupied by marine basins, are marked on a
general tectonic map or on a map of the basement surface. Shallower negative structures of higher elevation usually link up with these deep basins or
depressions. It is natural to assume in such cases that these shallower negative structures could have been convenient sites for the valleys of rivers
whose mouths were located on the shores of seas which had filled the vast
depression regions. The dependence of the hydrographic network on the
structural pattern can be observed on nearly all platforms of the world. It
is therefore possible to draw tentative schematic diagrams showing the location of the major ancient deltas. This type of prediction may be especially successful in regions which have not undergone major rearrangement.

Detection and tracing of ancient river valleys are best done on a regional
scale. Isopach maps should certainly be supplemented by lithofacies maps.
Usually the thickness of alluvial deposits increases gradually in the estuarine
parts of valleys, mainly due to the increase of sand formations in the section.
Bands of thicker terrigenous deposits in erosion valleys are usually oriented
perpendicular to the ancient shoreline.

A study of ancient alluvial-deltaic beds may be conducted according to
the following program: recognition based on facies assignment and other
features; distinction of individual facies (fluvial, floodplain, oxbow) and determination of their relationships; analysis of the thickness of sediments
deposited in river valleys and at paleoriver mouths; identification of basal
horizons of individual alluvial complexes; tracing of buried fluviatile formations both in section and in plan.

It should be recalled that sediments of many alluvial facies in the buried
state are not infrequently destroyed by erosion. In most cases the upper

parts of alluvial beds are removed; the basal horizon in stream channels is the best preserved. Sediments of the scour or the basal horizon are readily recognizable by the admixture of gravel-pebble material or by the presence of coarse sand fractions. Of the various types of alluvial deposits, fluviatile sediments are the most favorable for the accumulation of oil and gas. One should use all the available direct and indirect criteria in order to detect them.

Cross-bedding is a most characteristic feature of fluviatile formations. Its forms are closely related both to the mechanical composition of the sediment and to the variations of streamflow. In view of this circumstance the cross-bedding produced by river action differs significantly from other types of oblique bedding. Its basic genetic characteristic is sets of parallel cross-beds, mainly with a unidirectional slope of 25—30°. The thickness and extent of a set are dependent on the strength of the current. Cross-bedding is sometimes pointed up by coarse-grained fractions or by the detritus of vegetable and animal remains.

The bedding in the shoal zone adjacent to the river channel often becomes cross-wave-like or even imbricate. Gently sloping cross-bedded sets resulting from the accretion of sediments parallel to the slope of the shoal may also be encountered there. The thickness of cross-bedded sets usually ranges from 10 to 70 cm, rarely reaching 1 m. As a rule, fluviatile sets taper from bottom to top in section, thus distinguishing them from marine deposits which lack this feature.

Variation in grain size from coarse-grained sand to siltstone is possible both within individual beds and within sets. On bars, where the velocity of current is always higher, coarse material is deposited, while finer material is carried to the deeper portions of the river bed. Since the location of bars and reaches constantly changes, the grain-size distribution of the sedimentary material changes accordingly. Studies of fluviatile deposits according to textural and mechanical features must be supplemented by investigation of changes in the roundness, sorting and mineral composition of the grains. The general trend of the hydrographic network is determined from the decrease in the abundance of unstable minerals.

Sand deposits acquire a somewhat different aspect as one moves from the intradelta to the prodelta. In particular, layers with indistinct, gently undulating ripple bedding are encountered among the cross-beds. The internal structure is the result of the joint action of river and sea water. The foreset beds of the prodelta usually are gently sloping and predominantly show criss-cross-bedding.

A tentative forecast of the location of a paleodelta may sometimes be based on the fact that major present-day rivers are often inherited from older or even quite ancient ones. This inherited character is quite evident on the East European Platform, in West Siberia and North America, as well as in other regions of the world.

If one knows the position of channel sands relative to the shoreline of a marine paleobasin, one can easily picture the trend of hydrocarbon migration which is usually opposite to the monoclinal subsidence of the nearshore zone toward the center of the basin. Such bands should preferably be drilled according to a triangle system along the path of hydrocarbon migration. If the first wells do not discover an oil or gas pool, then, from the amount of hydrocarbon gases in the waters or from other components, a resultant line of the maximum of dissolved gases and other positive indications of oil and gas is determined in the system of the well triangle.  For instance, the following amounts of hydrocarbons were determined in the formation water of the the first bed tested (%): well 1, 0.3; well 2, 0.1; well 3, 0.6.  As follows from the graph (Figure 46), the gas or oil-and-gas pool in the given bed must be situated to the northeast of well 3.  The next bed is then perforated, and the same tests are repeated after the first bed has been carefully sealed off by cementing.  The resultant lines obtained from other geochemical and geophysical investigations are determined according to the same principle.  Analogous operations are carried out for all promising beds.

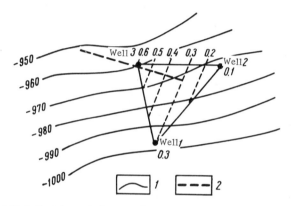

FIGURE 46.  Schematic diagram of well location in the triangle system used in prospecting for lithologic pools:

1 — contours in meters; 2 — resultant lines.  The numbers 0.2 to 0.6 designate the percentage of hydrocarbons in the formation water.

The resultant lines will apparently have different directions for different beds, especially if they refer to different horizons.  In such a case the direction is selected according to the data for one bed or to an average for several beds with resultant lines of approximately the same direction.

In the direction of the resultant lines or their mean azimuths the next two wells are drilled in the system of a new triangle whose base is the resultant line obtained and one of whose vertexes is the well of the previous triangle with the highest indexes or the point on the resultant line closest to this well.

Thus, well triangles are drilled until either pools are discovered or the boundaries of the area earmarked for exploration are reached.  It is advisable to complement the drilling with seismic prospecting and also ground or aerial radiometric prospecting of the territory.  In the process of studying gamma-ray logs and the radioactivity data on rocks and fluids, the direction of prospecting must be modified accordingly, taking into account the data on the dissolved gases, fluctuations in static levels and chemical properties of the waters, all of which are obtained in the course of this work.

This is only one of the possible variants of the method of detecting pools in lithologic traps.   There are also other alternatives, some of which are discussed by Maslov (1968).   The guiding principle in selecting the method of prospecting for lithologic pools must obviously be to proceed from the history of geological development of the region and the given area, knowledge of the paleogeographic conditions of formation of known or presumed producing formations, the regularities of zonal oil and gas accumulation, facies changes, the distribution of various lithologic varieties, and also from other prerequisites for the accumulation of hydrocarbons.

Part Two
# SOVIET EXPERIENCE

## 11. ATTEMPT AT RECONSTRUCTING THE PALEOGEOGRAPHIC DEPOSITIONAL ENVIRONMENT OF THE PRODUCTIVE TERRIGENOUS BEDS OF THE YASNAYA POLYANA SUBSTAGE IN THE VOLGA-URAL OIL-BEARING REGION

*Brief Lithofacies Description*

The terrigenous complex of the Ural-Volga region is assigned to the lower and middle parts of the Visean stage (Lower Carboniferous). We have studied the Middle Visean terrigenous beds of the Bobriki horizon and partly of the Tula horizon. We studied these beds not only in the Volga-Ural region, but also throughout the East European Platform, in order to obtain a more complete picture of their genetic features and depositional environments. The territory is composed mainly of argillaceous and sand-silt rock types not infrequently containing coal measures and lenses. Even though the lithology of the rocks is rather homogeneous, frequent facies changes are quite characteristic.

In the northwest of the East European Platform, on the eastern slope of the Baltic Shield, deposits of the Visean stage have been identified. These deposits have a kaolinite-bauxite member, 5—25 m thick, at the base and a ferruginous clay member, 0.5—23 m thick, at the top. These beds, which date back to Yasnaya Polyana times, are not generally continuous, being limited by crystalline outcrops of the Vetrenny belt in the north and by the Kanev projection of Precambrian rocks in the south. To the southeast and east they are overlain by younger rocks and they outcrop only on the western slope of the Urals.

The Yasnaya Polyana substage in the Timan-Pechora region includes terrigenous beds underlain by limestones of the Kizel horizon and overlain by mottled clays of the Aleksin horizon. It was not possible to distinguish the Bobriki and Tula horizons. The terrigenous beds consist of an alternation of sandstone, siltstone and carbonaceous mudstone. Limestone intercalations occur in the top part of the section. In the south of the Pechora Range there is a thick suite of so-called "grindstone" consisting of well-sorted quartzose sandstone cemented by oxidized petroleum and representing a giant ancient oil pool exposed at the surface. The thickness of the substage is variable, but generally increases from northwest to southeast where it attains 350—370 m.

The Kolva-Vishera margin of the Yasnaya Polyana substage consists of a rather uniform, mainly terrigenous series or coal-bearing suite which is not subdivided into individual horizons. This suite as a whole is comparable with deposits of the same age in the Kizel region. Chochia (1955) identified three types of rocks in it: western, represented by clays, sometimes carbonaceous shale alternating with sandstone and limestone intercalations;

central and eastern, represented mainly by deltaic sediments with lenticular coal inclusions; easternmost, characterized by inequigranular quartzose sandstone rich in carbonized plant detritus. The thickness of the substage increases from west to east (from 0—30 to 130—150 m).

In the Kizel region, which contains a well-studied coal basin of the same age, there is a terrigenous coal-bearing series of Early to Middle Visean age. The subdivision of the coal-bearing deposits in the basin is generally based on the lithology. The lower and upper argillaceous suites transitional from limestone to arenaceous-argillaceous rocks, or vice versa, can usually be distinguished. Sandwiched between these transitional suites are coal-bearing suites, mainly of argillaceous-arenaceous composition, and intermediate suites consisting almost completely of sandstone. There are varying estimates of the number of coal-bearing and intermediate suites throughout the basin. This is perfectly natural, since the depositional environment within a large subaerial delta, in which the Kizel coal basin was formed, could not have been the same throughout the territory. The overall thickness of the coal-bearing deposits in the central part of the basin ranges from 160 to 250 m, decreasing to 40—50 m to the west and north.

Rocks of this substage in the Kama region near Perm and in Udmurtia consist essentially of terrigenous deposits. Their lower boundary is usually drawn along the bottom of the sand layer located at the base of the Bobriki horizon or at the top of the mudstone member overlying the Tournaisian limestones. The upper boundary is essentially arbitrary and is generally taken to be the base of the first limestone interlayer containing Tula fauna. The section of the terrigenous series is characterized by an alternation of sandstone, siltstone and mudstone, among which carbonaceous varieties and coal also occur. The silty-clayey rocks are rich in plant detritus; paleosol beds with remains of root systems are quite common. Relatively sharp increases in thickness are usually due to the presence of cross-bedded quartzose sandstone and are generally noted in areas of erosion depressions or in regional troughs.

A thick complex of terrigenous sediments, in places showing signs of erosion and sharp angular unconformities, in places lacking any apparent erosion features, overlying Tournaisian or Famennian carbonate rocks, forms a broad zone in Tataria and the Kuibyshev region, especially in the Kama-Kinel depression. The lower portion of the complex is represented mainly by marine terrigenous deposits, the upper portion — by continental rocks. Pozner (1955) identified the independent Malinovka beds which were later incorporated into the Malinovka substage. The latter consists of well-sorted montmorillonite mudstones which give way upward to claystone-siltstone-sandstone series characterized by irregular horizontal bedding disturbed by detritovore burrows or root systems of plants. All this, together with the hydromicaceous composition of clays and the presence of the intraformational erosion, indicates that the Upper Malinovka beds underlying the Bobriki horizon were deposited both in the nearshore-marine zone of a shallow basin and on its shores.

More than 50% of the Bobriki horizon higher up in the section in the Kuibyshev Transvolga region consists of sandstones containing mudstone and siltstone members with intercalations and lenses of coal. The sandstones and siltstones are almost invariably quartzose, with a minor content of heavy minerals.   The mudstones, usually consisting of kaolinite-beidellite with a small content of hydromicas, are rich in carbonaceous detritus.   The intercalations of carbonaceous clays have a kaolinite composition.   Irregular horizontal bedding and cross-bedding are widely developed in the sandstones.   Not uncommon are inclusions of plant root remains vertically intersecting the bedding, as well as burrowing organisms, which are responsible for the mottled texture of the rock.   To the southeast of the area of maximum thickness of the Lower Carboniferous terrigenous series, it is gradually replaced by carbonate rocks.   The thickness of the series, together with the terrigenous portion of the Tula horizon, ranges from several meters on the Tatar arch to hundreds of meters in the Kama-Kinel depression.

In Bashkiria the lithology of the terrigenous series differs little from that of similar deposits in other Ural-Volga regions.   In places the Bobriki horizon appears to lack an hiatus at the top of the Tournaisian limestones.   The latter are sometimes almost totally eroded and in such cases the thickness of the terrigenous formations of the Yasnaya Polyana substage increases to 100 m or more owing to the sandstones.   Within the arenaceous-argillaceous complex the occurrence of coal is fairly common and thin limestone intercalations are rare.   The terrigenous series varies in thickness from 1 or 2 m in the south of Bashkiria to 40—60 m in the Birsk depression and to more than 100 m in its northwestern part.

In the lower reaches of the Volga, encompassing the Saratov and Volgograd regions, the terrigenous series has been identified both according to E-log data and to spore-and-pollen analysis.   The arenaceous-argillaceous series is made up of fine-grained (rarely medium-grained) quartzose sandstones which are particularly well developed on the right bank of the Volga. In the Transvolga region, the section mainly consists of argillaceous rocks and siltstone, in places with intercalations and thin layers of dolomite, sandstone or siderite.   The overall thickness of the Yasnaya Polyana terrigenous beds on the right bank increases toward the deepest portion of the Ryazan-Saratov depression, where it reaches 200 m.

The Middle Visean deposits of the Dnieper-Donets depression are of mixed terrigenous-carbonate type; to the northwest of the Donbas the limestones are gradually replaced by argillaceous-arenaceous deposits.   The thickness of the Lower and Middle Visean ranges from zero on the slopes of the Voronezh anteclise in the Ukrainian crystalline massif to 250 m or more in the depression proper.

The lithofacies features of the terrigenous series in the East European Platform make it possible to identify three major types of deposits: palustrine-lacustrine, alluvial-fluviatile-deltaic, and nearshore-marine. The first type is the most widely developed. Sediments of the second type, though not so widespread, play an important role in the structure of the series under study.

## Physiographic Environments at the Time of Deposition of the Productive Beds of the East European Platform

The lithofacies of a particular stratigraphic interval is the starting point for reconstructing the paleogeographic depositional environments corresponding to this interval.   To this end, lithofacies diagrams or maps are prepared which, together with isopach maps, depict the general features of the physiographic environment in the area being studied.   The degree of reliability and detail of the reconstruction depends on the available geologic information and its spread over the studied area.

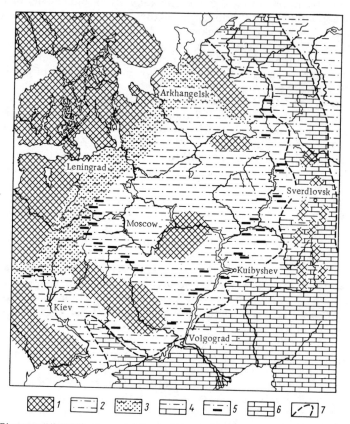

FIGURE 47.  Diagrammatic lithofacies map of the East European Platform during Yasnaya Polyana time:

1 — ancient land areas undergoing erosion; 2 — alternation of arenaceous-argillaceous rocks of palustrine-lacustrine and alluvial types; 3 — arenaceous-argillaceous diluvial-eluvial and lacustrine deposits; 4 — arenaceous-argillaceous nearshore-marine deposits; 5 — coal-bearing deposits; 6 — marine carbonates; 7 — postulated shoreline.

Figure 47 shows our lithofacies map of the Bobriki and Early Tula deposits of the East European Platform.   Analysis and refinement of this map in accordance with available knowledge of the platform (mainly according to borehole data) have made it possible to detect the following principal elements of the ancient physiography.

Regions with Raised Relief Forms

The following were the most elevated relief elements of the ancient East European Platform, which supplied detrital material to the depressions: the eastern and southeastern slopes of the Baltic Shield, the Voronezh and Ukrainian massifs, Timan and, partly, the Tokma and Upper Kama uplifts.

On the eastern slope of the Baltic Shield, in the region of the Onega River, the Lower Carboniferous is represented by red beds, the base of which consists exclusively of rudaceous conglomerates containing pebbles of basic and ultrabasic rocks. In the Andoma-Vytegra region there are traces of erosion at the contact between the Lower Carboniferous and the underlying horizons of the Upper Devonian. The Lower Carboniferous sediments, consisting of conglomerate, sand, siltstone, clay and limestone, extend over 25–30 km and are divided into three series.

The lowermost is the so-called Patrov arenaceous-argillaceous series which is in turn subdivided into three members: $a, b$, and $c$; it is evidently of Yasnaya Polyana age. The $a$ member overlies the eroded Upper Devonian and is made up of lagoonal-continental formations. In the north it is represented by mottled clays with an interlayer of dark refractory clays up to 60 cm thick. The thickness of the refractory clays increases gradually to the southwest and in the Vytegra region the overall thickness of the $a$ member reaches 10 m. Further southeast, in the Tikhvin region, bauxites make their appearance and the thickness increases to 25 m. Toward the south, in Borovichi, the bauxites disappear, their place being taken by coal lenticles and kaolin and plastic refractory clays.

The $b$ member is located higher up in the section. It consists mainly of sands divided into two distinct units. Its lower part consists of thick sands of variable structure and composition, showing the following features: lenticular bodies, commonly eroded, poorly sorted and inequigranular. Conglomerates and pebbles are encountered at the base. The bedding of the sands is variable, taking the form of sets of cross-beds with a southeasterly orientation. The angle of slope of the cross-beds ranges from 5 to 35°. The average thickness of the sets is 50 cm.

The upper part consists of fine-grained sands and siltstones with clay intercalations. The sands are fairly well sorted and lack prominent bedding. The overall thickness of the Lower Carboniferous sands in the Tikhvin region ranges from 4 to 40 m. A series of clays with sand interlayers is located higher up in the section.

Tournaisian deposits are generally absent in the northwestern regions of the East European Platform. According to N. S. Igolkina, Visean fauna was detected in the arenaceous-argillaceous series, suggesting that this section, especially the $a$ and $b$ members, should be correlated with the Yasnaya Polyana substage.

The character of sands overlying various Devonian horizons indicates that their formation was associated with the redeposition of terrigenous material derived from dry land. The dry land in question was made up of Devonian and older rocks and was situated to the west and northwest of Dvina Bay and the Onega and Ladoga lakes. This is confirmed by the mineral composition of the sands which consist mainly of stable minerals such as quartz, zircon, tourmaline, staurolite, etc.   Unstable minerals (feldspars, micas, hornblende and apatite) are much less common than in the Devonian deposits.

More recent studies carried out by Goryanskii et al. (1958) provide more accurate estimates of the age of the Lower Carboniferous arenaceous-argillaceous deposits which extend as a belt from the North Onega to the Tikhvin region.   The terrigenous series within these limits, having a thickness ranging from 4 to 40 m and directly overlying the Devonian sediments, is correlated with the Tula horizon according to spores and other plant remains.   In the lower part of the series mottled argillaceous members show facies variation to carbonaceous members.   The former are eluvial-diluvial, the latter lacustrine.   This is also borne out by their different locations within the pre-Carboniferous Devonian relief: mottled clays correspond to raised parts of the relief, while carbonaceous clays occur on lowlying relief elements.

The absence of deposits of the Tournaisian stage and of the Bobriki horizon further confirms that the northwestern part of the East European Platform underwent erosion between Frasnian and Tula times.

Consequently, during the Middle Visean there was a land area made up of arenaceous-argillaceous Devonian rocks which underwent intensive erosion and, in all probability, served as the principal source of terrigenous material for the rest of the East European Platform.

In most of Timan, as well as in the sections of the Kotlas, Yarensk and Bolshiye Porogi wells, the Yasnaya Polyana deposits are either totally absent or are only represented by thin Upper Tula layers.   This suggests that these regions should also be regarded as raised forms of the ancient relief. At the end of the Tournaisian—beginning of the Visean, the sea withdrew eastward and northeastward from the northern regions of the East European Platform.   The Timan region of that time was dry land sloping downward from west to east.

Other elevated portions of the East European Platform, aside from the eastern slope of the Baltic Shield and the western part of Timan, also experienced erosion at the beginning of the Visean.   The Voronezh Massif is one such region.   Lower Carboniferous rocks are absent in its upper part. On the southeastern slope of this massif, within the Saratov and Volgograd regions, the Upa, Cherepet' and Kizel horizons, whose deposits were apparently truncated by the pre-Visean erosion, are either partly or wholly missing from the Lower Carboniferous section.   In places where the Kizel horizon is preserved, it is represented either by carbonate or argillaceous varieties.   Silicified limestone and dolomite breccia and conglomerate predominate among the former, while the argillaceous sediments contain abundant plant detritus.

According to the data from deep wells drilled on the southwestern slope of the Voronezh Massif in the Novoaidar and Verkhnyaya Tarasovka regions, the Lower Carboniferous directly overlies the crystalline basement.

Therefore, during the period in question no sedimentation took place on a considerable part of the ancient Voronezh syneclise; this area underwent erosion.

The Ukrainian crystalline massif was also an elevated land area which was undergoing erosion.   The Visean and Namurian deposits there often directly overlie Precambrian crystalline rocks.   The Early Visean deposits are encountered only in localities where the steep slopes of the massifs grade into the gentler margins of the Dnieper-Donets depression.

The truncation of the crystalline rocks of the Ukrainian Shield and the transport of the erosion products to the Pripyat basin are corroborated by the section of the Yasnaya Polyana substage in the Belorussian Polessie. The sands of this substage are the distintegration products of the quartzites that are abundant in the Ovruch region. This also suggests that at the beginning of the Visean the Precambrian rocks in the north of the Ukrainian Shield were overlain by a sedimentary cover of younger formations. The greater part of the Ukrainian Shield was characterized by raised relief forms.

The territory of the Lithuanian-Belorussian Massif, which lacks Carboniferous deposits, was also an elevated land area at the beginning of the Visean. Its sedimentary cover underwent erosion, and the erosion products were carried into the Pripyat, Moscow and Lvov depressions.

Terrigenous material was, in particular, supplied to the Pripyat depression following the erosion of the igneous and sedimentary rock complexes, as indicated by the petrographic features of the arenaceous-argillaceous deposits. The region of the Tokma-Gorki arch rose somewhat above the generally flat landscape of the East European Platform. We include here areas adjacent to the Tokma arch within the 10-m isopach contour of the terrigenous beds of the Yasnaya Polyana substage. Deposits of this stratigraphic interval are totally absent in the northern part of the zone, near the town of Gorki.

Territories where terrigenous rocks are either very thin or absent formed plateau-like elevations of the ancient relief. In all probability, they were hilly or erosion-dissected land areas which served as drainage divides.

### Zones Transitional from Raised to Depressed Forms of the Ancient Relief

Nearly all gently sloping elevations of the East European Platform during the Early Visean graded into relatively depressed elements of the ancient relief via fairly broad transition zones in which the sediments of the terrigenous beds are poorly represented.

For instance, the Poretskoye test hole, which penetrated this zone between the Tokma arch and the Melekess depression, passed through a 9-m section of terrigenous rocks of the Yasnaya Polyana substage overlying an eroded surface of the Likhvin limestone. At the base of this section there is a 1.8-m clay member (dark, almost black, plastic, in places silty). The light fraction of the clays consists of 92.5% quartz and 5.7% feldspar. Higher up one encounters a light gray siltstone member giving way to a clay parting toward the top, the latter containing disseminated pyrite and pyritized burrows of detritovores. The terrigenous series is overlain by limestones of the Tula horizon.

To the northeast of the Poretskoye borehole, in the region of the Anish-Arya interfluve, the terrigenous series is 7—11 m thick. It consists of clay and mudstone alternating with sandstone and siltstone. Locally, the mudstone contains inclusions of carbonized vegetal remains and intercalations of carbonaceous sediments.

The Ust-Buzuluk area on the southeastern slope of the Voronezh Massif belongs to the transition zone.  Here the terrigenous beds of Yasnaya Polyana age pinch out toward the arch, but gradually become thicker in the direction of subsidence.  The deposits of the Yasnaya Polyana substage in the Ust-Buzuluk region overlie an erosion surface of Tournaisian carbonates.  The substage starts with a black carbonaceous clay member, 3—4 m thick.  Higher up it is replaced by bluish-gray clays, 4—5 m thick, containing carbonaceous remains and thin intercalations of light-gray limestone. These, in turn, are overlain by gray and dark gray clays, often with pyritized plant remains and accumulations of reworked macrofauna.  These clays contain silt material giving rise to bedding, as well as inclusions of pyrite, marcasite and siderite fragments.

The Yasnaya Polyana rocks show low resistivity (5—10 ohm · m) on the E-logs and positive values on the SP curves.  Several prominent peaks (30—50 ohm · m) stand out against a background of low resistivities, with negative SP maxima for limestone and sandstone interlayers.

The Ust-Buzuluk region is part of the uplifted portion of the southern slope of the Voronezh Massif.  The Carboniferous deposits there dip gently to the southeast.  The Yasnaya Polyana substage is 0—37 m thick.

Rocks of Yasnaya Polyana age penetrated by boreholes in the Ust-Buzuluk region indicate that, despite the presence of palustrine-lacustrine facies (thin intercalations of coal), the territory was a relatively elevated relief element where marshes and lakes often dried up and coal accumulation was insignificant.  This is also evident from the presence of fusain and fusain-xylain in the coal of the transition zones.  The formation of these materials can be attributed to the accessibility of oxygen during the transformation of plant tissues, owing to the relative dryness of the peat bogs.

The vast transition zone from elevated to depressed relief forms extended along the eastern and southeastern slopes of the Baltic Shield.  It flanked regions of older rocks which, being even more elevated elements of the ancient land area, underwent intensive erosion.  Deposits of these zones are characterized by small thickness and variable lithology and structure, both laterally and vertically.  Their formation is due mainly to ephemeral streams following heavy rains.  The enormous width of the transition zone, attaining hundreds of kilometers, is indicative of the exceptional flatness of even the elevated relief elements of the East European Platform during the period in question.

## Humid Plains

The greater part of the East European Platform during the Early Visean time was occupied by a lowland humid plain which was drained by an extensive river network.  This plain was not uniform throughout.  Its individual areas had specific features caused by the tectonic structure and pre-Visean truncation.  For instance, the deepest portion of the Middle Russian depression was occupied by the Moscow epicontinental basin; in the Pripyat depression or graben the basin was even deeper.  Rivers flowing down the Baltic and Ukrainian shields and the Lithuanian-Belorussian and Voronezh massifs debouched into such basins.  New larger rivers of the lowland type, which debouched into the seas fringing the platform, probably originated in marshes and lakes located in these basins.

The continental deposits of the humid plain may be divided into two main types, alluvial and palustrine-lacustrine, the latter predominating. Lakes, occupying depressions on the exposed surface of the Tournaisian limestone and fed by meteoric, river and subsurface waters, were filled with sediments and either disappeared or were overgrown by vegetation, developing into swamps and later into peat bogs. This is a rather rapid process due to constant supply of detrital material brought by the rivers. According to Slavyanov (1948), the complete infilling and overgrowing of Lake Baden will be completed within 12,500 years, and that of Lake Geneva, within 45,000 years. The same author cites as examples 118 lakes in the Tyrol which disappeared or developed into bogs during a single century.

The period of infilling and swamping of the lakes in the deepest portions of the East European Platform was marked by intensive accumulation of coal and the peneplanation of the relief. Swamps were formed not only on the sites of former lakes. The geomorphological features of a territory, its tectonic regime, the pedologic-geologic conditions, the vegetation and other factors are the main prerequisites for the formation of swamps. An essential condition for the development of vegetation, which later turns into coal, is excess humidity. In other words, atmospheric precipitation must exceed evaporation.

The numerous remains of woody and other terrestrial plants, as well as spores and pollen, in the Yasnaya Polyana substage suggest that the vegetation was of the humid tropical type. The most widely developed were lycopsids: lepidodendrons, sigillarias, etc.

The process of transformation of peat bog into humic coal took place under varying geomorphological conditions: in river valleys and on the plain, within platforms and on the seashore, in water divides and deltaic areas. In all cases, however, the source material was the remains of terrestrial swamp and aquatic plants. The genetic features and petrographic composition of coal are indicative of the conditions in which it was formed, and, consequently, of the physiographic environment in which it accumulated. According to Naumova (1940), the characteristic features of coal formed in limnic (lacustrine) environments are the presence of algae, the uniform distribution of macerals, and the absence of fusainized woody elements. Coals of the paludal (palustrine) type, on the other hand, are characterized by the absence of algae and the usual irregular distribution of macerals, and consist mainly of the remains of higher plants.

Sapropel coals containing algal remains, which are typical of lakes, are quite rare in deposits of the Yasnaya Polyana substage. The humic nature of most of the coal points to its predominant swamp origin.

Sediments of the limnic-paludal type are represented mainly by argillaceous, silt-sand, and carbonaceous-argillaceous rocks with coal beds and interlayers of varying thickness and with varying ash contents. The most common argillaceous rocks are mudstones rich in plant remains. Their composition — predominance of kaolinite and an insignificant admixture of hydromicas — serves as further evidence of the continental environment of their origin.

The sand deposits consist almost entirely of quartz with minor amounts of feldspar, mica and heavy minerals, pointing to strong chemical weathering in catchment areas which fed regions of the lowland plain. This humid plain sloped toward the sea and in some places developed into everglades.

The alluvial deposits of river valleys also occupy vast territories and play a major role in the structure of the terrigenous beds. They are represented by argillaceous-arenaceous rocks, among which sandstones and siltstones predominate. These show dislocated bedding or cross-bedding, usually occur as lenses and have varying thickness.

The mineral composition of the arenaceous rocks is rather homogeneous — mainly well-sorted and rounded quartz grains. Variations in grain size are noted, the 0.25—0.1 mm fraction predominating. The most typical heavy minerals are zircon (invariably predominant), rutile, tourmaline, epidote, kyanite and staurolite. It is sufficient to mention that the zircon content in the heavy fraction of the terrigenous beds of the Yasnaya Polyana substage reaches 25% or more. This suggests that the material supplied to the Pripyat depression was derived from relatively nearby source regions.

From a comparison of numerous sections of the Lower Carboniferous terrigenous beds of the East European Platform, one notes that the average grain size of the sand varieties increases somewhat in a northwesterly direction, being accompanied by a higher content of unstable minerals such as feldspar, apatite, hornblende and pyroxene.

The quantitative relationship between arenaceous and argillaceous rock types in river valley deposits is fairly variable. In some cases sandstones account for 25—30% of the rocks, while in others the percentage is as high as 70—75%. Intercalations, beds and lenses of coal are not uncommon in this type of sediments. It is difficult to recognize and distinguish between lacustrine-palustrine and alluvial-fluviatile deposits because of their intermittence, lack of persistence, and facies changes over small distances. An almost monomineralic composition of the terrigenous sediments indicates prolonged transport and, probably, repeated redeposition of detrital material which had undergone chemical weathering on elevated elements of the ancient relief and was transported by streams and deposited on the lowland plain. Much of this material was supplied to marine basins.

Deposits of the humid plain generally possess features typical of both lacustrine-palustrine and alluvial sedimentary environments. The well-known Moscow basin serves as an example. At the beginning of the Visean it was a lowland covered by marshes and relict or newly formed lakes with brackish or fresh water. Numerous streams and large rivers descending from the hills fringing the basin debouched into these water bodies. Rivers of the present-day type, such as the Svir, Neva, Volkhov, etc., issued from some of the lakes. These rivers flowed across a lowland covered with lakes and marshes, changing their courses, washing away peat bogs and redepositing enormous amounts of terrigenous material; they built up an alluvial plain within the platform. Widespread on the plain were peat bogs which were transformed into coal after having been buried under younger sediments. The genetic features of the fairly diverse coal deposits in the Moscow basin are most indicative of the paleogeographic conditions at the time of their accumulation.

Naumova (1940) identified the following facies on the basis of the various genetic features of the coal: profundal zone with boghead coal; sublittoral zone with cannel coal; littoral zone responsible for the accumulation of clarodurain coals with algal inclusions.  The three facies groups correspond to different zones of lakes.

In addition to lakes, on the more elevated relief elements there were also numerous marshes with plant associations that gave rise to the formation of purely humic coals.  The following marsh facies environments are distinguished: quagmires where semilustrous and semidull coals consisting of clarain and clarodurain were formed; partly flooded marshes where silky semidull coals (attritus) originated; marshes fed by springs, with dull durain coals; and dry peat bogs where fusain coal varieties accumulated.

Judging from the petrographic composition of the coals from several Ural-Volga regions, similar facies environments also existed in other areas of the Early Visean plain.  However, fewer boreholes have been sunk in these areas than in the Moscow basin, and it is therefore much more difficult to determine the environments in which their coal deposits were formed.

Deposits of the Bobriki horizon are rich in pyrite almost throughout. Strakhov (1959) believes that, owing to the high content of organic matter buried in the terrigenous sediments, the authigenic mineral forms of iron underwent transformation, giving rise to the more intensive development of pyrite.  This is why the Lower Visean coal-bearing rocks are characterized by high pyrite contents.  The fact that both marine and continental deposits show pronounced pyritization points to the salinity of the water bodies on the lowland plain and to their considerable content of sulfates.

Regions of the plain which were situated along the platform margins and bordered directly on the sea had certain distinctive features.  Such regions constitute the nearshore-marine lowlands whose shores occasionally became marine shoals favorable for the development of quagmires and mangrove thickets.  The presence of limestone and dolomite interlayers is a typical feature of nearshore-marine deposits.  The sandstones in such regions are generally inequigranular and show different degrees of sorting.  According to Strakhov et al. (1959), the poor sorting is due to the uniqueness of the seacoast which at that time consisted of numerous islands surrounded by shoals and overgrown with trees.

Coasts of this type were characterized by stagnant waters, and the clastic material supplied from the shore was deposited immediately, scarcely undergoing sorting.

The occurrence of coal in such a type of nearshore-marine lowland differs from that in the alluvial plains and deltaic zones.  Although the content of plant detritus in the rocks is very high, and intercalations of carbonaceous mudstones and siltstones are very common, coal occurs fairly rarely, and as thin beds.

Most interesting in all respects are the nearshore-marine areas where alluvial-deltaic sediments were deposited.  These areas occupied extensive territories of broad estuarine valleys of rivers which debouched into the sea.  They had the form of dry land dissected into numerous islands by distributaries and channels.  It is here, in the zone of transition from dry land to sea, of interaction between river and sea, and of upbuilding and destruction that the age-long struggle between water and solid matter is most pronounced and the complex transformation of matter into materials with new properties is most intensive.

### Paleorivers

Humid plains are generally characterized by a gradual accumulation of sedimentary deposits.  However, in individual sections, usually extending as long strips, continental beds accumulate rapidly.  Their thickness increases regularly as the strips gradually widen.  This phenomenon is readily seen on the isopach map of the terrigenous beds of the Yasnaya Polyana substage within the East European Platform (Figure 48).

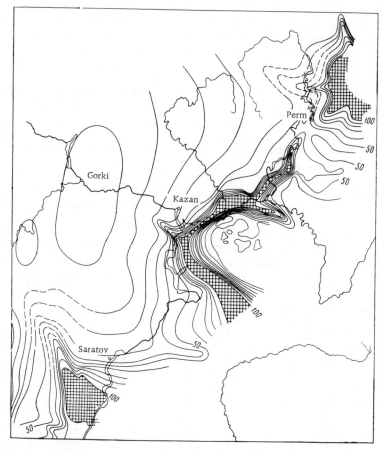

FIGURE 48.  Isopach map of the terrigenous beds of the Yasnaya Polyana substage in the southeast of the East European Platform.  Areas where the thickness of the beds exceeds 100 m are cross-hatched.

Thicker strips generally consist of alluvial arenaceous-argillaceous rocks associated with depressed elements of the structural plan of the crystalline basement and are formations of ancient river valleys of erosion-tectonic origin.  Almost invariably they have a high content of sandstones and silt-stones and the argillaceous rocks usually have a high content of coal.  Deltaic sediments are widespread in the estuarine parts of the valleys.  Numerous coal deposits and major carboniferous oil deposits are associated with them.  The rivers carried into the continental water bodies and seas not only terrigenous material, but also organic matter and solutions of mineral salts necessary for the normal development of the plant and animal kingdoms.  The rivers built vast alluvial plains on the land area and deposited thick deltaic formations in the coastal and shelf regions.  The formation and distribution of highly permeable rocks suitable for the migration and accumulation of oil and gas depended largely on river activity.

The main trend of the well-developed hydrographic network of the Middle Visean rivers, both large and small, was controlled by the tectonic structure of the platform, while their hydrogeologic regime was generally dependent on the climatic factor.  The abundant subtropical or even tropical rains and showers which fell on the East European Platform at that time fed the rivers and changed their levels and rates of flow.

The hydrodynamic force of the rivers and the intensity of denudation were controlled by the amount of meteoric precipitation.  Hence, the great significance of the hydrographic network in the erosion, transport and deposition of terrigenous sediments.  To date, several large valleys and deltas of paleorivers have been recognized with varying degrees of reliability, depending on the availability of factual data.  The names given to these ancient features are derived from their present geographic location.

*Timan-Pechora Paleoriver*

The Timan-Pechora paleoriver is situated in the northeast of the East European Platform and is readily discerned on the isopach map and on the map of the Yasnaya Polyana substage.  The Edzhid-Kyrtyn coal deposit is located in the mouth zone of the valley of this paleoriver (the right bank of the middle reaches of the present-day Pechora).  The coal beds assigned to the Visean stage are similar to those in the section of the Kizel coal basin, 550 km further south.  The section through the coal-bearing member, occurring within thick lenticular sandstone, contains pyrite concretions; rare siderite formations are also encountered.  Typical features of the arenaceous-argillaceous beds of the Edzhid-Kyrtyn region are frequent interfingering and rapid changes in lithology along the strike.

The most common grain size of the arenaceous rocks of the Timan-Pechora region is that of the silt fraction (0.1—0.01 mm).  The fraction larger than 0.1 mm predominates in sections of the Malaya Kozhva and Voi-Soplyas river basins; rocks with grain size exceeding 0.25 mm account for a considerable part.  The arenaceous-argillaceous rocks consist of rounded, less commonly angular-rounded quartz grains, the large grain-size fraction displaying the best roundness.  Angular grains predominate in the siltstone.  The light fraction of the sandstone contains 93—100% quartz, 1—2% feldspar, 1—2.5% muscovite, and up to 1% chalcedony.  The heavy fraction, rarely exceeding 1—2%, consists of pyrite, siderite, iron hydroxide, leucoxene, rutile, tourmaline, zircon and other minerals.

The inequigranular, medium-grained sandstones in the western regions on the left bank of the Pechora are usually lenticular. In some places arenaceous rocks account for about half of the entire section. Quite common are bituminous and oil-bearing sandstones alternating with clay and mudstone members. In the eastern regions of the right bank of the Pechora, mudstones locally containing remains of marine fauna predominate. These data confirm the presence of alluvial-fluviatile deposits of a large plain paleoriver in the Pechora depression. Thick formations of sandstones impregnated with oxidized petroleum are exposed in the quarry of a grindstone plant at the mouth of the Voi River and along the Bolshoi Soplyas. They consist almost exclusively of quartz (up to 99% of the light fraction). In the heavy fraction, which rarely exceeds 2% of the sandstone material, authigenic ore minerals predominate, while zircon, rutile, garnet, tourmaline, and some other minerals are present in small amounts.

FIGURE 49. Walls of the grindstone quarry — sandstones impregnated with oxidized petroleum — at the mouth of the Voi River in the Timan-Pechora region (a) and closeup of this exposure (b) (photograph by N.I. Markovskii)

The grain size of the sandstone varies both in section and plan. It
changes from coarse-grained to silt size within a single member or even
within a single lens. The predominant fraction is 0.25—0.1 mm. The sandstones on the right bank of the Pechora, located further to the east, are more
fine-grained than those on the left bank and contain a considerable admixture of the silt-size material (0.01 mm).

Lenticular intercalations of inequigranular and medium-grained sandstones are also encountered. They are generally well rounded and well
sorted. The larger the grain size, the rounder the grains. Angular grains
are present only in silty varieties. The porosity of the sandstones ranges
from 3 to 13%.

Cross-bedding, with the inclination of the laminae varying from 25 to 30°,
is very common in the sandstones at various levels of the grindstone quarry
(Figure 49) where 40—45 m of the section are exposed. The cross-bedded
sets dip to the northeast and east. Regular horizontal bedding was encountered only in the sandstone (0.5 m) at the base of the quarry's southern wall,
being indicative of tranquil conditions of sedimentation. This area must
have been occupied by a bay or lagoon, which was not influenced by currents
and waves, prior to the advance of the delta toward the marine basin.

Irregularly bedded lenticular sandstone varieties showing a general eastward dip occur higher up in the section of the same wall. Further up one
observes a markedly undulating intraformational erosional unconformity
between silty and overlying coarse-grained sandstones. The western part
of the quarry's northern wall is made up of more homogeneous sandstones
which become somewhat argillaceous and finely bedded toward the east.

An inspection of the quarry suggests that the current in the fairly narrow
distributary was strongest between the northern and southern walls. The
pinching out of individual lenticular sand members is quite evident. Obviously, it is difficult to judge the entire deltaic zone from a stretch of
150—200 m (the length of the quarry), but even the exposed portion of the
section gives us a very vivid picture of the depositional environment that
once existed there. The visible thickness of the producing formation in the
quarry does not exceed 50 m, and that in the exposure along the Soplyas
River is 125 m.

The principal distinctive feature of the sandstones in the region of the
Voi and Bolshoi Soplyas rivers is their impregnation with oxidized oil which
accounts for up to 8—10% of the total weight of the rock.

On the right bank of the Pechora, 50—60 km south of the Voi River, is
located the Edzhid-Kyrtyn coal basin where the Visean producing formation
attains a thickness of 300 m. Here the section is similar to that to the south
of the Kizel paleodelta. The geologic structure and occurrence of coal in
the Edzhid-Kyrtyn basin have been described by K. G. Voikovskii-Kriger,
G. G. Bogdanovich and others. Coal deposits were also discovered in the
Kyrta-Yola, Troitsko-Pechorsk and other regions.

The producing formation can be traced in exposures and wells in the
basins of the Podcherem, Ilych and other rivers, as well as 45 km north of
the Shchugor River mouth, where its maximum thickness attains 370 m.
Further northeast, in the upper reaches of the Usa River and in the basin
of its left-bank tributaries (Kozhim and Kos'yu), the terrigenous beds are
predominantly argillaceous and overlie the Tournaisian limestones without

any visible hiatus.  They are represented mainly by shales with abundant siderite inclusions.  In some places the shales are carbonaceous, slightly bituminous or ochreous, usually having a dark, almost black color, and are thinly bedded or foliated.  The siderite occurs as concretions or lenses. It commonly forms pseudomorphs after fauna or algae.  Goniatites of Visean age have been found in the siderite concretions.  The thickness of the argillaceous-siderite series here ranges from 250 to 300 m.

Siderite is known to be a mineral which is formed in a reducing environment with a deficiency of oxygen and containing decomposing organic matter. Its presence in the studied section suggests that conditions favorable for the formation of source beds existed in the relatively deep-water coastal zone, in the vicinity of the thick deltaic deposits of arenaceous sediments.  Siderite deposits are also encountered on the right bank of the Pechora, in exposures along the Bolshoi Paton, Podcherem and Vuktyl rivers.

The terrigenous beds located to the west of the above-mentioned regions are of an evidently continental sediment.  They show facies changes and often contain coal.  Sandstones of alluvial origin account for a considerable part of the section.  Inequigranular and medium-grained sandstones occur as a distinct, relatively narrow strip extending from southwest to northeast through the Vuktyl, Voi, Yugid, Kyrta-Iol and Malaya Kozhva regions.  In the western regions of the Pechora basin there occur petroliferous sandstones which in places alternate with coal-bearing argillaceous members and carbonaceous mudstones.  This productive complex gives way to clayey-siderite and then carbonate deposits as one proceeds further east or northeast.

Most researchers who have studied this region are of the opinion that the detrital material was derived from the east.  We consider this assumption to be erroneous.  The mineral composition of the sandstones, together with the increase in their thickness from west to east and the successive facies changes, is indicative of the direction of transport.  The sandstones are generally monomineralic, consisting essentially of quartz.  Predominant in the heavy fraction are authigenic and rather stable minerals which survive repeated redeposition of the sediments.  Such material could only have been derived from a dry-land source area covered by sedimentary rocks; this description exactly fits the greater part of the East European Platform and the Timan region at that time.  The land surface on the opposite sea shore was made up of a metamorphic rock complex which supplied clastics of polymictic composition, as exemplified by the Lower Carboniferous deposits on the eastern slope of the Urals.

Dobrotvorskaya (1947) claimed that the material was derived from the west.  She based her assumption on the increase in the average grain size of the sandstones in a westerly direction.  However, she did not rule out the possibility that some clay material could also have been brought from the east.  Dobrotvorskaya rightly stated that the beds were formed under nearshore-marine and nearshore-continental deltaic conditions.

The frequent interfingering of siltstones and sandstones, their lenticular occurrence, the irregular bedding or cross-bedding, the variable thickness, as well as the rather variable proportion of arenaceous and argillaceous rocks in the section of the left-bank regions in the middle reaches of the Pechora, all point to their deposition in a deltaic zone which passed into the gently sloping shelf of the marine basin toward the east and northeast.

The terrigenous, mainly arenaceous-argillaceous beds attain maximum thickness (up to 300 m) in the region extending along the middle reaches of the Pechora (from the town of Pechora in the north to the settlement of Kyrt in the south).  It is here that the alluvial-deltaic deposits of the Timan-Pechora paleoriver pass into prodelta deposits.  One should also not rule out the possible presence of sand formations of the bar type in this zone. In the northeastern part of the Pechora syneclise, the Yasnaya Polyana substage is represented by marine carbonates.

*Kizel Paleoriver*

The discussion here relates not so much to the river itself, as to its delta which has been studied in numerous exposures and mines in the Kizel coal basin (investigations by D. V. Nalivkin, I. I. Gorskii, N. S. Gordetskaya, P. V. Vasil'ev, G. Ya. Zhitomirov, G. A. Smirnov, N. G. Sazhin, I. V. Pakhomov and others).

The rather well-studied Lower Carboniferous producing formation virtually lacks paleontological evidence; the meager finds of fauna make it possible to date this series as not older than the Chernyshino substage and not younger than the Yasnaya Polyana substage.  The assignment of the greater part of the terrigenous beds to the Middle Visean, based on both the fauna and the spore-and-pollen complex, is fairly well grounded.  However, the age of the lower portion of the section, which in many places overlies an erosion surface of older rocks, remains unknown.

The accumulation of coal-bearing deposits need not necessarily be due to oscillatory movements.  Considerable differences in sedimentation, even within a relatively small area and during the same time interval, may be due to the ruggedness of the relief, frequent changes in the hydrodynamic regime of the rivers, determined by the alternation of periods of intensive rainfall and aridity, and to other factors.  The complex sequence of terrigenous formations in the Kizel basin consists of members in which certain lithofacies or rock types predominate, corresponding to specific paleogeographic environments.  This is the main clue to their distribution pattern.  For instance, the distribution of clastic material in zones of present-day deltas is governed by the intensity of flow in the distributaries and channels. V. P. Baturin has convincingly demonstrated the significance of the dynamics of the sedimentation environment, unrelated to oscillatory movements.

The structure of the coal-bearing beds in Kizel and similar regions is determined mainly by ancient relief forms.  River valleys passed through lowlying regions where the peak accumulation of sand material occurred. Terraces and raised interfluves were sites where marshes developed and peat accumulated.  In zones intermediate between fluviatile and palustrine facies there are transitional or mixed facies of oxbows, lakes, blind creeks, etc.  All these factors are responsible for the diversity of facies in many coal-bearing areas.

The general facies environment in alluvial-deltaic regions is characterized by a number of local facies changes which complicate the structure and make it difficult to study the producing formations.  This is why it is not easy to correlate different sections even over small distances.

The terrigenous beds in the Kizel basin are made up of relatively few rock types which sometimes occur repeatedly in the section. These are mainly sandstones, siltstones, mudstones, and coals.

The sandstones accounting for the major part of the Bobriki horizon are typically oligomictic. They occur as lenses and show cross-bedding and frequent transitions of certain varieties into others. We shall mention the following sandstone varieties which are widespread in the basin:

**Coarse-grained sandstones** have a light fraction composed mainly of quartz grains 0.25–1 mm in size, although they also contain gritstone varieties with grain size from 2 to 5 mm, as well as fragments of carbonized wood. Most of the grains are rounded or oval and those larger than 1 mm are often split or show cracks and scratches. Heavy minerals are either poorly represented or totally absent. The cement is of the basal, overgrowth or agglutination type. It commonly consists of siderite nonuniformly distributed throughout the rock.

**Medium-grained sandstones** are represented by well-rounded and well-sorted quartz grains, 0.25–0.50 mm in size. The degree of roundness decreases on transition to finer-grained sandstones. Heavy minerals are represented by small amounts of zircon and tourmaline, their percentage increasing somewhat in the finer-grained varieties. In isolated cases flakes of mica and grains of apatite and anatase are encountered. Secondary quartz, less commonly, siderite or chalcedony constitute the cement.

**Fine-grained sandstones** are the most widespread. They consist essentially of well-rounded and well-sorted quartz grains, 0.25–0.1 mm in size. The grains are usually elongated or irregular in shape, less commonly oval.

**Very fine-grained sandstones** with grain size 0.05–0.1 mm are most commonly found in coal-bearing members or layers underlying and overlying the Bobriki horizon. Quartzitic varieties predominate; sandstones with argillaceous-ferruginous and siderite cement are less common. In places the coal-bearing suites contain pyritized sandstone, less commonly varieties with calcareous cement.

**Siltstones** (0.01–0.05 mm) contain, in addition to quartz, admixtures of mica and accessory minerals such as zircon and tourmaline, as well as occasional grains of apatite and garnet. The quartz grains have an irregular angular shape, indicative of their transport in suspension. The cement is mainly quartzose-clayey, less commonly quartzose-ferruginous or clayey. Siltstones are widespread in coal-bearing deposits where they occur as individual layers or thin intercalations alternating with fine-grained sandstones, the transition from variety to variety being scarcely discernible.

**Mudstones** play a minor role in the terrigenous beds of the coal-bearing formation in the Kizel basin. They usually contain silty material (15–25%) and carbonized plant detritus, occasionally flakes of muscovite. Accessory minerals are represented by zircon and tourmaline.

The coal that is mined in the basin is classified as humic and, in places, sapropel-humic. Some 20 coal beds and interlayers have been discovered. Only three or four of them are of sufficient thickness to merit mining. According to its outer appearance, the coal ranges from dull to bright. On the whole, the coal is poorly vitrinized material and characterized by a high content of spores, reaching 60% in the dull coals but averaging 25%. The coal is of medium rank, ranging from long-flame to fat.* The moisture content is 2.3—5.8%; ash, 19—31%; sulfur, 3—6.4%; volatiles, 37—48%.

A study of the section through the coal-bearing deposits reveals the frequent alternation of rocks of all types, both vertically and laterally. This makes it difficult to establish the distribution pattern of the various deltaic formations. It was, nevertheless, possible to ascertain that the number of coal beds (more specifically, the amount of lenticular coal bodies) within the section, although independent of the thickness of the coal-bearing horizon, is controlled by the facies environments of sedimentation. A higher coal content generally occurs in the argillaceous rocks, while the sandstone either contains little coal or none at all. It was estimated that the Bobriki horizon consists, on the average, of 65% sandstone, 26% siltstone, 7.6% mudstone and 1.4% coal. The limiting values (%) for these rocks in the section are: sandstone, 39—90%; siltstone, 4—54%; mudstone, 0.3—49%.

Consequently, sandstone is the principal constituent of the producing formation in this basin. Its granulometric composition is as follows: 1.0— —0.5 mm fraction, 0.02%; 0.5—0.25 mm fraction, 6.25%; 0.25—0.1 mm fraction, 80.75%; 0.1—0.01 mm fraction, 12.35%; less than 0.01 mm fraction, 0.63%. If one compares these figures with the mechanical composition of ancient and present-day deltaic and fluviatile sands, one readily sees that the Kizel sandstone differs little from typical sandstones of fluviatile-deltaic origin. The maximum abundance of sandstone (up to 80—90%) was noted in the southern and southeastern parts of the Kizel basin. Its role also increases somewhat in central regions. Areas with 45—49% sandstone are less common and occur mainly in the northwestern part of the basin. It is rather difficult to establish any definite distribution pattern for the sandstones because of their lack of persistence. Of special significance among the sand deposits of the coal-bearing formation is the so-called "roof" sandstone overlying the major coal-bearing suite and developed almost throughout. Its average thickness in the basin is 30 m, the limiting values being 4 and 80 m; in areas where the roof sandstone shows increased or maximum thickness the coal content decreases or coal beds are totally absent. On the other hand, in localities where this sandstone becomes thinner interlayers of coal are more numerous. For instance, areas such as the northern part of the Kospash-Poludino syncline, the northern part of the Shumikha syneclise and the Usva field, where the roof sandstone is from 70 to 145 m thick, are either very poor or completely lacking in coal. It seems natural to conclude that the areas where sand deposits predominate and the coal content drops markedly coincide with the main watercourse or with the distributaries of the ancient delta, while areas with thinner sandstone beds and a higher content of coal represent swamped islands in the deltaic zone. In the former sand material accumulated in the dynamic medium of streamflows, while in the latter peat bogs were formed in relatively tranquil environments.

---

* [According to the Russian classification, long-flame coal is of rank 1, while fat coal is of rank 3.]

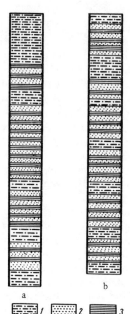

a       b

FIGURE 50. Comparison of sections through the Volga delta (a — exposure of the Bolda River channel) and the lower portion of the Kalinin mine (2nd horizon) in the Kizel coal basin (b):

1 — arenaceous-argillaceous rocks; 2 — sand and sandstone; 3 — clay and mudstone.

The eastern continuation of the Kizel coal-beàring deposits is limited by truncation; westward they lie at considerable depths and have hardly been explored to date. Only a relatively narrow area (10—15 km), extending meridionally for approximately 200 km, has been studied in detail. This is by no means the entire original area of Early Visean coal accumulation in the ancient deltaic zone and adjacent lowland.

The genetic features of the major commercial coal seams in the Kizel basin confirm the complex facies environment in the deltaic zone where coal accumulation was taking place. Three main zones, coexistent but differing in conditions of sedimentation, may be identified on the basis of the distribution of individual rock types, including coal, both in section and in plan.

1. Zone of relatively stable islands on which swamps were formed, later developing into peat bogs. This latitudinal zone of fairly large deltaic islands was situated mainly in the northern part of the Kizel alluvial-deltaic lowland where the producing formation has a maximum coal content.

2. Unstable zone, or zone of running waters (the main course of the river, its largest distributaries and channels). This zone was often formed and preserved for a long period of time in the southern areas of the delta. It is here that the dynamic medium ruled out the possibility of coal accumulation, and sand was virtually the only sediment to be deposited.

3. Mixed zone characterized by frequent changes of facies environment. This zone is encountered almost throughout and is rather variable both in vertical section and in plan. In such zones the alluvial islands and river

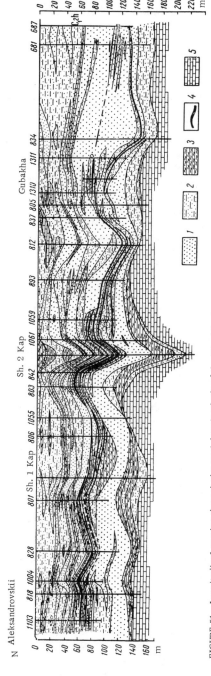

FIGURE 51.   Longitudinal section through the coal-bearing beds of the Kizel basin:

a – sandstone;  2 – siltstone;  3 – mudstone;  4 – coal;  5 – limestone.

banks were often inundated or breached by temporary channels and meandering river courses, became swamped or dried up, and then again underwent certain changes.   Although the number of coal seams in the deposits
of this zone may be considerable, they are usually thin, have a high ash content, and commonly pinch out or are replaced by carbonaceous-argillaceous
rocks.

The main physiographic zones identified within the deltaic region are not
particularly persistent, but they sometimes display certain inherited features
or repetition of the same depositional conditions over a considerable time
interval.   On the whole, their location, persistence and recurrence were dependent on the changing destructive and constructive action of river and sea
waters.   In some places islands and coastal shoals were formed, while in
others they were eroded to be built up once again, and so on.   All this resulted in a peculiar pattern of alternating rock types in the section of the
terrigenous beds.   The longitudinal N—S section through the Kizel region
(Figure 51) serves as an excellent illustration of this alternation.   In constructing the section, the top of the terrigenous beds, corresponding to the
base of the Tula limestones, was taken as the datum horizon.   This made it
possible to eliminate the effect of the present-day structural pattern and to
reconstruct partially the relief which had existed prior to the deposition of
the coal-bearing sediments.

Sets of cross-bedded sandstones (Figure 52) attaining several meters in
thickness are known to occur in various locations in the Kizel basin;  they
are usually typical of sandstone members occurring in the middle of the
section.   The cross-bedding, generally diagonal, takes the form of alternating sandstones of different  grain size showing parallel bedding with a
primary dip of 20—30°, sometimes reaching 40°.   A pronounced angular unconformity usually separates the cross-bedded sets from the overlying and
underlying sandstones which have more regular bedding.   This type of
cross-bedding, which is typical of deltaic deposits, is very common in the
Kizel basin.   Less common is another pattern in which regular bedding at
the base of a member gradually changes to cross-bedding toward the top.
This is accompanied by an increase in the thickness of the deposits.   The
contact with the overlying beds usually takes the form of an angular unconformity.

Measurements of the spatial orientation of the cross-beds in sandstones
of the coal-bearing series furnish a rather important characteristic.   These
measurements, carried out in various mines and natural exposures, show
that the direction of the primary dip of the beds is remarkably constant.
According to all available data, the primary dips of the cross-beds are predominantly to the east-northeast and east-southeast.   In less than 30% of
the cases the dips are to the north-northeast, and only in a few isolated
instances — to the northwest.   Numerous measurements of cross-bedding in
mines of the Kizel region also show that the primary dip of the beds was
either to the northeast or almost easterly.

Smirnov and Svirshevskii (1955) distinguished the following distribution
pattern and features of the cross-bedding.   The cross-bedded sets in the
center of the Kizel basin display the greatest thickness and steep primary
dips to the east and northeast.   Twenty kilometers to the south, in the valley

of the Usva River, where the cross-bedded sandstones are less thick and
the beds have a gentler dip, the orientation is no longer as constant as in
the central part of the basin.   Further south (45—50 km), in the basin of the
Chusovaya River, the sandstones lack cross-bedding.

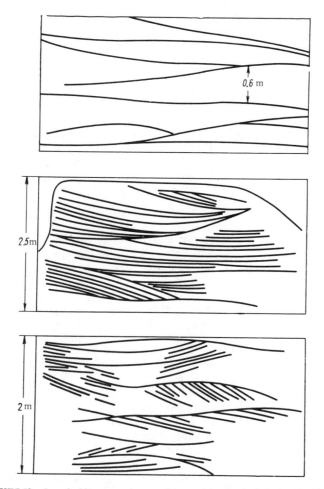

FIGURE 52.   Cross-bedding in sandstones of the coal-bearing beds in the Kizel basin

Consequently, the cross-bedding, which is quite pronounced in the central
part of the basin and has a specific azimuthal orientation, indicates the direction of flow of the river which built the Kizel delta.   This direction was
from west to east, i. e., away from the East European Platform.   The mouth
of this paleoriver was located approximately on the Kospash-Gremyachinsk
line; the main course was situated to the west of this line, in the latitudinal
depression between the Upper Kama arch in the north and the Bashkir arch
in the south.

Another ancient stream channel may exist in the northern part of the basin, as suggested by the isopachs in the Vishera region. This has yet to be verified. However, the existence of several closely spaced river mouths is compatible with their general distribution pattern on the seashore where they are sometimes absent and sometimes almost continuous. The regular decrease in thickness of the terrigenous beds toward the west, accompanied by a gradual decline in the role of the sandstones, and the absence of coal and argillaceous interlayers in the easternmost parts of the Kolva-Vishera region led N. G. Chochia to conclude that continental conditions existed in these regions and that shallow marine facies predominated further westward. This viewpoint is definitely erroneous. Those who are of the opinion that the Kizel paleoriver flowed from the east rather than from the west were guided by similar reasoning. The increase in the thickness and sandstone content of the Lower Carboniferous terrigenous beds was not due to the proximity of the source area, as believed by Chochia and others, but should rather be attributed to the fact that the rivers transporting this material flowed in this direction and that maximum deposition took place in the river mouths.

### Paleo-Kama

The paleo-Kama is one of the largest rivers of Middle Visean age and its deposits are well preserved in the Kama-Kinel depression system, in the Bira depression, and partly on the southern dome of the Tatar anticline. The mouth of this paleoriver was situated southeast of the Samara Bend, where its valley attained a width of almost 150 km. The distribution pattern of the thick alluvial formations deposited by this river shows up on the isopach map.

There are varying opinions as to the age and environment of deposition of the thick Lower Carboniferous terrigenous series which forms an elongated strip in the Transvolga region near Kuibyshev, in Tataria and in adjacent regions of Bashkiria and South Udmurtia. Its stratigraphic division was discussed earlier in this book. Most researchers regard the lower mudstone portion of the terrigenous sediments as an independent unit (Malinovka beds) made up of a regressive complex of marine deposits giving way to continental sediments in the upper portions of the section.

The upper part of this series, corresponding to the Bobriki horizon and partly to the Tula horizon, is composed of sandstone, sand and siltstone with subordinate intercalations of carbonaceous clay and mudstone, as well as coal. The sand-silt varieties are monomineralic (quartz), have variable thickness, show irregular bedding and cross-bedding and are rich in vegetable detritus, root systems and traces of burrowing organisms.

As far back as 1955 we suggested that the coal-bearing beds passing through the saddle between the northern and southern domes of the Tatar anticline and filling the Melekess depression are of erosion-tectonic origin.

Other opinions also exist. Khachatryan (1957), Troepol'skii et al. (1957) and Lobov (1960) believe that the depression was produced by tectonic inversion. Kleshchev et al. (1957) and Aver'yanov (1960) put forward the hypothesis of facies replacement of the terrigenous beds by a synchronous carbonate formation (Devonian-Tournaisian stratigraphic interval). Filippov (1957) and Grachevskii (1959) think that the depression is a buried, intraformational, aggradational topographic feature inherited from an un-compensated Upper Devonian depression.

We regard the Malinovka beds, which have a relatively limited distribu-tion and are of local significance only, as deposits of a relict water body which had been preserved since Tournaisian time and existed within the depression as a bay until the developing delta of the paleo-Kama displaced it to the southeast. This is borne out, in particular, by the presence of mixed Tournaisian-Visean fauna in the Malinovka beds, as well as by traces of temporary marine ingressions into the river paleovalley.

What indicates that the Yasnaya Polyana terrigenous beds are of fluviatile origin? Firstly, the shape of the beds themselves, extending as a rather narrow strip over a considerable distance; moreover, as a rule, increase in thickness due to sand deposits is accompanied by increased width of the valley. The action of rivers tends to level the relief. A relative pene-planation of the paleorelief within the strip in question took place as far back as the first half of the Tula. The Upper Tula transgression contributed to the leveling of the territory. If the marine limestones of the Tula horizon overlying the terrigenous beds are taken as the datum horizon, and the thick-ness of the terrigenous beds is plotted with reference to it, it is then possible to determine the shapes of the ancient relief forms which were generally filled by fluviatile deposits. The lenticular occurrence of the sandstones and the fact that they pinch out toward the margins of the paleovalley suggest their fluviatile origin.

Secondly, the composition of the rocks in the coal-bearing portion of the section, represented by sandstone, sand and siltstone alternating with oc-casional thin layers of clay, mudstone and carbonaceous shale or coal, also suggests their alluvial character. These rocks are usually rich in vegetal detritus; remains of root systems and traces of burrowing organisms are not uncommon. The sandstone in core samples sometimes displays cross-bedding, the slope of the beds reaching 22—25°. The carbonaceous mudstone is often pyritized.

The monomineralic composition of the clastic material (essentially quartz) is a rather typical feature of the terrigenous rocks. The heavy minerals include zircon (10—15%), rutile (4—6%), tourmaline (1.5—3%) and epidote (1—2%). This composition, taken together with the fact that the clastic ma-terial is fairly well rounded and well sorted, suggests a remote provenance, prolonged river transport, and, probably, also repeated redeposition.

In the upper reaches of the paleo-Kama in the Malinovka area the large upper part of the thick terrigenous series is definitely of alluvial origin. The arenaceous-argillaceous rock complex here contains coal, carbonaceous mudstone and thin intercalations of siderite. The chemical composition of the latter is: 39.72% FeO; 3.88% $Al_2O_3$; 7.35% CaO; 3.75% MgO; undeter-mined 9.88%, loss on ignition 27.76%.

The siltstone locally displays truncated cross-bedding. Sandstone is absent in the lower part of the section, and the terrigenous series consists of clay which is usually dark-gray, microlaminated, compact, in places silty, with interlayers of fine-grained dark-gray dolomite and siderite concretions.

The chemical composition of the siderite and siderite-dolomite formations is indicative of their accumulation in semisaline stagnant water bodies, such as lagoons, oxbows and floodplain lakes. This is also evident from the wide-spread pyritization of the argillaceous rocks.

The sections of other wells (at Malinovka and in the neighboring areas of Gorkoovrazhnaya, Radayevskaya and Rakovskaya) are analogous to that described above. The two last-mentioned areas are located on opposite slopes of the ancient valley whose width here is approximately 30 km. The predominance of sandstone in the coal-bearing portion of the section and its specific features point to floodplain and fluvial environments of deposition. For instance, the thickness of the lenticular deposits ranges from 1.5 to 35 m. The mineral composition of the sandstone (essentially quartz with occasional stable heavy minerals) remains constant irrespective of the thickness of the lenses and their position relative to other rocks; it is also independent of grain size. The number and size of the sand lenses in the paleovalley differ considerably from area to area. Their grain size is also not constant, this variability being probably due both to differences in the rate of flow and to changes of the water course.

Higher up in the paleovalley the deposits of the Yasnaya Polyana substage were penetrated by the Melekess test hole (depth interval of 1478—−1595 m). Here the Bobriki and Tula horizons were identified; their thicknesses are, respectively, 91.5 and 25.5 m. The substage consists mainly of sand-silt rocks, with subordinate intercalations of carbonaceous clay and mudstone, carbonaceous shale and coal.

The Melekess borehole penetrated five coal beds (in all probability, lenticular deposits), 0.40—1.25 m thick, as well as several thin coal interlayers. It should be pointed out that coal is fairly widespread throughout the strip in which the thickness of the Yasnaya Polyana substage is great. It was detected by drilling in the Pokrovskaya, Malinovka, Radayevskaya, Rakovskaya, Gorkoovrazhskaya, Elabuzhskaya and Sarailinskaya areas and in several localities in Bashkiria. In contrast to the lowland plain, where coal usually occurs as a single thick bed, the section of the strip under study invariably contains several thinner coal beds. This serves as additional proof of the fact that this strip was part of a river valley.

Both the presence of coal and the mineral composition of the clays attest to the continental deposition of the Bobriki horizon in the paleovalley of the Kama. Several clay samples from this horizon, obtained from the Malinovka 20 well and the Radayevskaya 1 well, were studied by X-ray-structural, optical and other methods at VNIGNI (All-Union Petroleum Scientific Research Institute of Geological Exploration). They were found to be kaolinite with an admixture of kaolinized hydromica. This type of clay is known to be characteristic of continental (fluvial, lacustrine, etc.) facies.

In which direction did the paleo-Kama flow and where were its upper reaches?

The shape of the isopachs and the strip-like form of the thick alluvial deposits suggest that the upper reaches of the river were divided into two sources or resulted from the confluence of two rivers. One source, probably the major one, issued from a local depression in the vicinity of the town of Osa; its direction of flow was to the southwest, along the course of the present-day Kama. In the area where the Belaya River joins the Kama, the above-mentioned source merged with the second source flowing from the Birsk saddle. The paleo-Kama then flowed almost due west, and somewhat to the south of the present-day mouth it turned southeast, being fed by

additional tributaries along the way.  One of these tributaries, which flowed
down the Tokma uplift, is readily discernible on the right bank of the paleoriver between Kazan and Ulyanovsk.

There is nothing surprising in the fact that the paleo-Kama originated in
a swamped lowland area close to the seashore and flowed away from the sea,
rather than toward it.  The present-day Niger River shows similar features.
It also has its origin near the seashore, then describes an inland arc and
finally debouches into the Gulf of Guinea, forming a vast delta.  This direction of streamflow is apparently connected with the tectonic structure of the
river basins.  There must have been a reason for the paleo-Kama to flow
around the southern dome of the Tatar anticline; one gains the impression
that it was squeezed into the Birsk saddle between the Tatar and Bashkir
anticlines.

Descriptions of the section of the Lower Carboniferous Bashkirian
series, terrigenous in particular on the Birsk saddle, have been provided
by Vissarionova (1959) and other authors.  It was noted that the Malinovka
horizon does not occur in Bashkiria, whereas the Bobriki horizon shows
several typical sections.  The one of interest to us here is that associated
with the upper reaches of the river valley, or the so-called Kaltasy type
(after A. Ya. Vissarionova).  This type includes the middle portion of the
section through the terrigenous rocks occurring in Northern Bashkiria.  A
limestone interlayer at the top of the terrigenous member of Tournaisian
age is considered to be the lower boundary of the Bobriki horizon.  The
base of the lowermost limestone interlayer containing Tula fauna is regarded as the top of this horizon.

The following are typical features of the coal-bearing deposits in the
Arlan, Kaltasy, Yugomashev and Chegmagush regions: alternation of mudstone, siltstone, sandstone, carbonaceous shale and coal; total absence of
carbonate interlayers; confinement of sandstone to the lower and upper
parts of the horizon; predominance of carbonaceous-argillaceous rocks in
the bottom half of the section.  The average thickness of the Bobriki horizon
in the Birsk depression is 25—30 m.  However, it increases toward the north,
attaining 57 m in the Arlan area (well 39).  The rocks become predominantly
argillaceous toward the south, west and east; this is accompanied by a decrease in thickness and abundance of the sandstone.

Coal in the form of lenses and lenticular interlayers (a few centimeters
to 20—24 m thick) is usually encountered in the western and northern parts
of the Birsk saddle.  Dull (durain) and semidull (spore-durain) varieties
predominate; semibright (clarain-durain) and bright (clarain) coal also
occurs.  Most of the coal is rich in sporinite, thus indicating a flow of water
through the peat-bogs.  The humic coal of Bashkiria is very similar in its
properties to coal of the same age found in other Ural-Volga regions.

The coal deposits in the northwestern part of Bashkiria form relatively
narrow zones extending from south to north.  Most of the lenticular sandstones containing commercial oil are associated with these zones.  The
sandstone consists of fine-grained, rarely medium-grained, well-sorted
quartz cemented by the clay, carbonaceous-clay or clay-carbonate material.
The cross section through the Arlan zone (Figure 53) vividly illustrates the
character of the coal-bearing series.  It is a typical section of an alluvial

river valley, which differs from the estuarine part in that it has a single river bed and much thinner deposits.

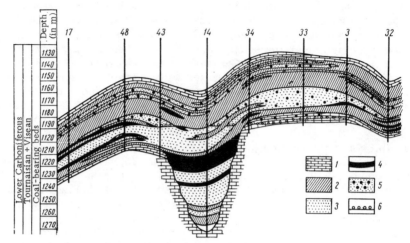

FIGURE 53.   Productive part of the section of the Arlan oil deposit:

1 — limestone;  2 — mudstone;  3 — sandstone;  4 — coal;  5 — petroliferous sandstone;  6 — water-oil contact.

The terrigenous beds in the Bashkirian part of the Cis-Ural depression are no more than 3—4 m thick and are highly calcareous.  On the western slope of the Urals they are represented by limestones.

The coal-bearing deposits gradually pinch out in the southern closure of the Birsk saddle.

The upper reaches of the paleoriver, extending from southern areas of the Perm region, have been poorly covered by drilling.  However, the available data show that here the coal-bearing beds consist of an alternation of light-colored sandstone and sand, siltstone, mudstone and clay which contain abundant carbonized plant detritus, with root systems and also coal lenses. The maximum thickness of the coal-bearing deposits here amounts to 55 m.

Thus, the alluvial-deltaic character of the paleo-Kama formations has been established at various points along its course (virtually from the source to the delta).  Further deepening and expansion of the river valley are noted in the direction of flow.  The terrigenous material resulting from the erosion of the previously deposited sedimentary formations was transported in the same direction.

### Shkapovo Paleoriver

Another relatively small river of Yasnaya Polyana age was identified in southwestern Bashkiria south of the Birsk depression, beyond a rather narrow drainage divide.  This identification is based on the distribution of the sandstone within the terrigenous coal-bearing beds whose thickness ranges from 8 to 30 m.  The sandstone usually occurs in the middle of the section and is represented by rather well-sorted fine- and medium-grained quartz

varieties.  They occur among argillaceous rocks, in the form of a single
thick layer and two to three interlayers separated by siltstones or clay-silt
deposits.  Sandstone is usually absent when the thickness of the terrigenous
beds drops to less than 10 m, but accounts for a considerable if not major
portion of the section when the thickness exceeds 10—15 m.

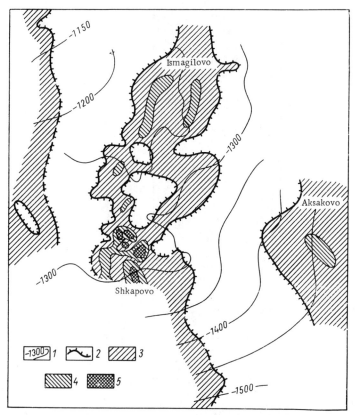

FIGURE 54.  Distribution of sandstone in the coal-bearing beds of southwestern Bashkiria (after Khalimov, 1960):

1 — countour lines of top of coal-bearing beds;  2 — pinch-out line of sandstone;  sandstone thickness:  3) from
0 to 10 m; 4) 10 to 20 m; 5) exceeding 20 m.

According to Khalimov (1960), the sandstone shows a specific distribution pattern in plan.  It forms narrow strips usually extending meridionally
over tens of kilometers (Figure 54).  Khalimov distinguishes three such
strips in southwestern Bashkiria: western (Stakhanovo-Znamenskoye),
central (Belebei-Shkapovo), and eastern (Aldar), separated by zones within
which the coal-bearing beds are made up of mudstone and impermeable
clayey siltstone.

The western sand strip is 3—4 km wide and its average thickness is 4 m.
The eastern strip has been covered by several wells; here the sandstone is
up to 5 m thick.  The central strip, extending N—S for 70 km across five oil

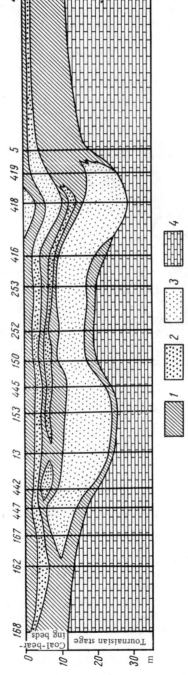

FIGURE 55.  Paleostructural cross section of the coal-bearing beds in the Shkapovo oil deposit  (after Khalimov, 1960):

1 — mudstone;  2 — siltstone;  3 — sandstone;  4 — limestone.

prospects, has been well studied (data from more than 500 wells).  The thickness of this sandstone strip, embedded in layers of mudstone and silt- stone, increases from 3—4 m in the north to 15—20 m in the south (the Shkapovo area); this increase is not uniform.  Its width gradually increases in the same direction, from 6 to 7 km in the Usen-Ivanovskoye area to 12—15 km in the Shpakovo area.  Figure 55 shows a cross section of this strip.

According to Khalimov, the strip distribution pattern of the sandstone, its sinuous outlines in plan, the lenticular shape of the sand body and the abrupt changes in grain size suggest that there is some connection between the sand accumulations and undercurrents.  This opinion is shared by Vissarionova (1959) who associates the accumulation of sandstones in the Birsk depression with marine undercurrents.

However, the vast amount of available factual material permits us to re- gard the strips extending across the Shkapovo and other regions of south- western Bashkiria as the deposits of one of the numerous paleorivers which flowed down the southeastern slope of the platform into the Middle Visean sea that then occupied the Caspian depression.  The prodelta of this probab- ly small river extended some distance to the south of Shkapovo, approximate- ly to the latitude of the town of Sterlibashev.

*Ryazan-Saratov Paleoriver*

The terrigenous rock complex in the Ryazan-Saratov depression, espe- cially in its southeastern part which has been best covered by wells, is variable both in lithofacies and thickness.  These beds generally increase in thickness from northwest to southeast and east, mainly due to the more frequent occurrence of sandstone and siltstone in the section.

Five major sand reservoirs have been distinguished in the Volga region near Volgograd.  One of them belongs to the Bobriki horizon, and the others to the Tula horizon.  These sand strata are not persistent; they often inter- finger with clays and pinch out.  They are usually lenticular in shape.  Both the Bobriki horizon and the lower part of the Tula horizon consist of clay, mudstone, siltstone, sand and sandstone.  Intercalations of carbonaceous rocks and coal are not uncommon.  The upper half of the Tula horizon con- sists mainly of carbonate rocks.

The strip of terrigenous beds of increased thickness in the axial portion of the Ryazan-Saratov depression broadens markedly on approaching the Volga, measuring more than 200 km across.  The greater part of the oil- and gas-bearing areas of the Volga region near Volgograd and Saratov is located in the rather broad near-Volga zone of the terrigenous beds.  The maximum thickness of the Yasnaya Polyana substage, about 200 m, was noted in the Ilovlya area.

Medium-grained sandstone with an admixture of coarse-grained material is often encountered at the base of the Bobriki horizon.  This sandstone overlies the eroded surface of the Malinovka beds.  The sandstone becomes finer higher up in the section and is sometimes replaced by sandy siltstone or even clay.  The sandstone is essentially monomineralic (quartz), with rare grains of feldspar, tourmaline and zircon.  The upper half of the sand bed contains a layer of inequigranular sandstone which gives way upward to fine-grained and silt fractions.

There are four principal sand beds in the lower terrigenous part of the Tula horizon. The first of them $(T_4)$ is a channel deposit which extends sublatitudinally. This is the main oil-producing bed in the Tula horizon of the Bakhmetyevka and Zhirnovsk fields. Further north the bed continues into the Saratov region. In other directions its facies changes to very compact argillaceous siltstones and clays. In the Zhirnovsk and Klenovka areas the bed is represented by a massive deposit whose effective thickness reaches 15—20 m, whereas in the Bakhmetyevka area it is split up into three reservoir layers separated by clays.

The second bed of the Tula horizon $(T_3)$ has been traced from the villages of Klenovka and Peskovatka in the north to the latitude of Archeda and Lipovka in the south. It continues as a narrow channel in the direction of Kachalino and Sirotinskaya. Its effective thickness reaches 14 m, in places 20 and 36 m. The maximum thickness is encountered in the Zhirnovsk-Tersk area adjoining the region of the Saratov dislocations.

The third sand bed $(T_2)$ is of limited extent and occurs in some regions in the form of lenses or irregular bodies. At the top of this bed the sandstone is sometimes replaced by limestone, and locally it interfingers with clay or siltstone. In the Linevo area the bed consists of medium-grained well-sorted sandstone. In the Korobkovo area it is fine-grained. Toward the southeast the sandstone is replaced by clay and limestone. The effective thickness of the bed is 0—9 m.

The fourth bed $(T_1)$ occurs in the upper part of the Tula horizon within an argillaceous member confined between two limestone beds. Its effective thickness is 0—5 m.

The sand bodies filling the northeastern channel system of the Ryazan-Saratov paleoriver have been traced by Gostintsev (1967) on the right bank of the Saratov and in the Transvolga region. The most productive oil and gas deposits with the largest reserves occur in structures associated with the longitudinal axis of the distributary channels, south of the Yelshan-Sergiyevsk and Sovetskoye-Stepnoye natural levees. The deterioration of the reservoir properties of the sandstone toward the margins of the ancient channels has a considerable effect on well productivity. Commercial accumulations of oil and gas are controlled by the intricate development of the sandstone as sinuous strips, 1—2.5 km wide, in a number of areas of the Volga region near Volgograd and Saratov.

The lithofacies of the terrigenous beds of the Yasnaya Polyana substage, the composition and shape of the sand bodies, as well as the presence of coal in core samples from the Ryazan-Saratov depression, suggest that these sediments are floodplain and fluviatile formations of a vast river valley. The upper reaches of the paleoriver were apparently located in the region of the swamped Moscow basin whence the paleoriver flowed southeast, receiving smaller tributaries on its way.

The river delta occupied the right bank of the present-day Volga, and its main course debouched into the marine basin approximately in the Kamyshin area. The system of minor distributaries and channels stretched from near Saratov to Volgograd. Drilling continues to reveal additional channel sands characterized by an alternation of various grain-size fractions showing nonhorizontal, generally oblique bedding and other features indicating their obviously fluviatile origin.

Borehole sections in the Bakhmetievka, Zhirnovo, Linevo and other areas situated in the intradelta region show an alternation of sandstone, siltstone and mudstone. The latter are micaceous in places or contain pyrite, and are rich in carbonized vegetal detritus almost throughout. Carbonaceous mudstone and coal are not uncommon; they usually pinch out over short distances.

An analysis of the coal has shown that it is humic, less commonly humic-sapropelic, transitional from brown to hard coal, and is classed as duro-clarain according to its microcomponents. The moisture content in the coal ranges from 1.07 to 14.58%, ash content from 4.17 (Linevo) to 35.87% (Bakh-metievka), sulfur from 2 to 6%; the volatile components of the flammable material constitute, on the average, about 50%. The relatively high content of volatile material is due to the abundance of spores, cuticles and resinous re-mains. More than 50 coal occurrences were detected notwithstanding the rather limited core sampling of test holes. All this bears witness to the fact that the greater part of the Volga right bank was occupied by the intradelta of a large plain river.

The grain size of the sandstone and siltstone is quite variable, owing to changes in the rate of flow. However, the fraction from 0.1 to 0.25 mm pre-dominates, its content varying from 40 to 97% within individual layers. In order to compare the sandstones of the terrigenous beds in question, which occur in certain river paleovalleys, with sands of present-day rivers flowing in plains, their mechanical compositions are listed in Table 6.

TABLE 6. Mechanical composition of sandstones of river paleovalleys and present-day river sands

| No. of well or sample | Depth (m) | Area or sampling site | Content (%) of fraction, mm | | | |
|---|---|---|---|---|---|---|
| | | | 1–0.5 | 0.5–0.25 | 0.25–0.1 | less than 0.1 mm |
| 25 | 1072–1079 | Bakhmetyevka .................... | — | 0.7 | 84.0 | 14.2 |
| 23 | 1155–1161 | Zhirnoye ......................... | 0.1 | 4.5 | 85.2 | 5.4 |
| 4 | 1275–1280 | Linevo ....... ................... | 1.0 | 6.7 | 84.8 | 5.1 |
| 12 | 1770–1776 | Korobki ......................... | — | 6.9 | 91.1 | 2.0 |
| 17 | 1817–1820 | Ilovlya ......................... | 9.0 | 9.0 | 87.1 | 3.9 |
| 3 | 2487–2492 | Umet ............................ | — | — | 93.3 | 4.6 |
| 1 | 1516–1527 | Melekess ........................ | — | 77.9 | 11.4 | 9.0 |
| 20 | 1421–1425 | Malinovka ....................... | — | 85.8 | 9.3 | 4.9 |
| According to many samples | | Kizel paleoriver ................ | 0.02 | 6.25 | 80.7 | 13.0 |
| 276 | Edge of bank | Volga near Saratov .............. | 0.1 | 0.5 | 91.0 | 7.4 |
| 277 | The same | Volga near Volgograd ........... | — | 11.0 | 82.8 | 6.2 |
| 278 | " " | Don near Kalach ................. | 0.7 | 1.0 | 95.8 | 2.5 |
| 279 | " " | Northern Donets near Bolshaya Kalitva. | 1.7 | 1.6 | 85.1 | 11.4 |
| According to many samples | | Dnieper near Kherson ............ | 0.5 | 31.5 | 68.0 | — |

Table 6 shows that the sandstones of the Ryazan-Saratov paleoriver differ little from those of the Kizel paleodelta or from sands of present-day rivers flowing in plains (Volga, Don and Northern Donets). All of them consist largely of the fractions from 0.25 to 0.1mm. This points to the similar power of the streamflow (rivers) that transported them. However, the sandstone of the paleovalley in the Melekess depression differs somewhat in that the coarser fraction is more widespread. Apparently, the river had a stronger flow in this location, as is also suggested by the deeper erosional downcutting of its bed.

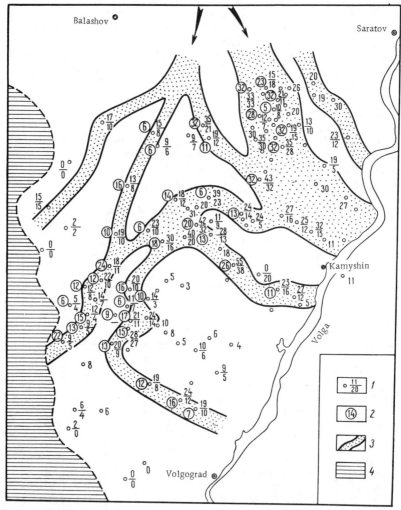

FIGURE 56. Distribution of channel sands in the paleodelta of Yasnaya Polyana age in the Volga region near Volgograd:

1 — wells (the total thickness of the sandstone (in meters) is given in the numerator, and the effective thickness in the denominator); 2 — average porosity according to geophysical data; 3 — channel sands; 4 — zone lacking sandstone.

In contrast to the sandstones in the river valley of the ancient Kama, those of the Ryazan-Saratov paleoriver contain feldspar, which is indicative of the proximity of the provenance. The granites and their disintegration products exposed in the anticlinal part of the Voronezh anteclise could have constituted such a source.

The mouth of this river and in particular of its delta were delineated more accurately in the course of oil exploration and drilling. Fairly well delineated at present are a part of the major course and several distributaries with lenticular channel sands. Figure 56 shows that the main course of the river, which apparently existed from the end of the Tournaisian till Late Tula times, cut its course from northwest to southeast.

In the Bobriki horizon alone this channel left a strip of sand 30—60 m thick and more than 30 km wide. On both sides of the main channel there occur channel sands whose width also measures several kilometers. The sand strips have the appearance of an "intrusion" into predominantly argillaceous rocks. In the central part these strips are made up of medium-grained sands that are almost totally devoid of clay admixtures. Toward the slopes fine-grained sands with a considerable content of the clay material make their appearance. Near the margins of the strips the sandstone is almost entirely replaced by clay.

The Bobriki sandstone consists essentially of quartz (87—97%). The principal heavy minerals are zircon (19.8—43.2%) and tourmaline (17.7—32.6%). Also present are rutile (5.7—16.8%), garnet (0.2—0.9%), brookite and anatase (0.9—2.6%), epidote (up to 1.3%), glauconite (0.1—6.1%), very rarely kyanite (up to 0.5%), and corundum (up to 0.6%). The ore minerals are represented by 4.6—18.3% pyrite, 17.2—35% leucoxene, and 1.4—6.8% ilmenite.

The siltstone of this zone is characterized mainly by a psammitic-silty texture. It contains 65—95% sand-silt material and 10—35% cement. The mudstone has imbricate and foliate-imbricate structures, the scales and fibers of the clay minerals having the same optic orientation. They also contain numerous carbonized fragments of vegetal tissues which also lie parallel to the bedding plane.

All these data attest to the deltaic nature of the terrigenous formations in the Bobriki horizon.

*Pripyat-Dnieper Paleoriver*

Another paleoriver of Yasnaya Polyana age has been detected in the southwestern part of the East European Platform. This river had its source in the lakes and marshes of the Pripyat basin and debouched into the Dnieper-Donets marine gulf. The estuarine part of this river valley was situated northwest of Poltava. The above assumption is based on the following evidence:

thickness variation both of the entire Yasnaya Polyana substage and of its individual terrigenous and carbonate complexes;

replacement in the northwesterly direction of apparently marine rocks widespread in the southeastern part of the Dnieper-Donets depression by the increasingly continental deposits including thick lenses of sandstone, carbonaceous mudstone and even intercalations of coal;

abrupt change in thickness over a small distance across the general strike of the terrigenous sediments, which is observed on the Kolaidinetsky and other territories and is due to the increase in the amount of sandstone of presumably fluviatile origin.

The petrographic-mineralogic association of rock-forming minerals in the Dnieper-Donets depression is similar to that of the Pripyat downwarp, the only difference being the lesser amount of the unstable, readily disintegrating minerals in the southeastern part of the depression. All these data indicate the presence of a paleoriver whose delta has to be delineated more accurately with the aid of data from boreholes to be drilled on the territory where it is assumed to occur.

The lithofacies and thickness of the Middle Visean deposits, as well as the nature of their coal occurrences in this part of the platform, suggest the following paleogeographic environments of their formation.

The rivers which transported vast amounts of terrigenous material from the Belorussian-Lithuanian, Ukrainian and, partly, Voronezh regional uplifts debouched into the rather deep Pripyat basin. The vigorous action of rivers in this region prevented the accumulation of coal on a broad scale.

Within this water-rich basin, there must have been numerous deep fresh- and brackish-water lakes rather than swamps, this being confirmed by the thick sand deposits. The Pripyat-Dnieper paleoriver must have had its origin in a fairly large lake and debouched into the Dnieper-Donets marine gulf. There are grounds for believing that this paleoriver was not very long and had an abundant flow like that of present-day rivers such as the Volkhov, Svir, Narva or Neva.

There is no doubt that, in addition to the Pripyat-Dnieper paleoriver, other streams also flowed from the Voronezh and the Ukrainian crystalline massifs directly into the sea.

## Marine Basins

We have repeatedly mentioned the transition from obviously continental deposits of Middle Visean age to nearshore-marine and marine facies. Although this transition is not always equally distinct, its confinement to the eastern and southern boundaries of the East European Platform is beyond doubt. The marine analogues of the Yasnaya Polyana substage can be traced along the western slope of the Urals and in some localities in the Caspian syneclise, and are readily distinguished in the Dnieper-Donets depression. They are represented by sediments of estuaries, bays and shallow seas which were probably linked by deeper-water regions of the common marine basin. Available data enable one to identify individual regions occupied by the sea or by somewhat desalinated marine gulfs.

### Timan-Pechora Gulf

In the Timan-Pechora province, northeast of the Kozhva basin, the deposits of the period under study consist almost exclusively of carbonate sediments, which also extend further north, up to the present-day shores of the Pechora and Kara seas. The limestone and dolomite occurring here point to the existence of a marine basin whose boundaries shifted in the course of time.

According to Kas'yanov (1949), intercalations of pure limestone are encountered on the right bank of the Pechora River, in the region of the Van-
gyr, in the upper reaches of the Bolshaya Synya and even on the Bolshaya
Patoka River. At times marine conditions also prevailed in the left-bank
regions of the Pechora, e. g., along the courses of the Voi-Vozh, Malaya
Kozhva, Yugid, Voya and Soplyas, where interlayers of carbonate rocks also
occur.

In the southeastern regions of the Middle Pechora basin the Yasnaya
Polyana substage contains several layers of calcareous mudstone with
siderite intercalations. Siderite deposits are most widespread in the basin
of the Kos-Yu River and its right-bank tributaries where they are of economic interest. Over the rest of the territory the siderite formations form
a fairly narrow strip running along the boundary of the carbonate sediments.

An environment favorable for the diagenetic formation of commercial
siderite deposits existed in the regions of fresh-water lakes, semisaline inland seas, and gulfs containing muds rich in organic matter, often interbedded with siltstone (Strakhov, 1948). However, such environments could
also have existed in normal marine basins with small amounts of carbonate
sediments.

L. V. Dobrotvorskaya has reported that beds with marine microfauna are
encountered in argillaceous members of the Lower Carboniferous producing
formation only on the right bank of the Pechora River, whereas no marine
fauna was found on its left bank. Fairly rich fauna of the Lower Visean type
is known to occur in the mudstones and siderite concretions of the Kyzhim
region. The above data may be considered sufficient proof of the fact that
the territory west of the Middle Pechora was formerly dry land, while that
to its east was a marine area.

To date numerous coal occurrences and individual coal deposits have been
discovered in the Lower Carboniferous terrigenous beds in the southeastern
part of the Komi ASSR (in the upper reaches of the Vychegda, in the
basin of the upper reaches of the Izhma, in the basins of the Veslyana, Vishera and Ilych, as well as in the upper reaches of the Pechora). It follows,
therefore, that south of Yedzhid-Kyrta a broad band of terrestrial coal-
bearing sediments extends nearly as far as Kizel; nearshore-marine formations occur to the east of this band. These formations were to a considerable extent incorporated in the fold zone of Western and Central Urals
and were eroded. This is why the zone of transition from the ancient humid
plain to the marine paleobasin is not distinct everywhere.

*The Ural Sea*

The Ural sea must have been a marine strait in the middle part of the
Ural geosyncline, situated between larger seas in the north and south.

The Yasnaya Polyana substage on the Kolvo-Vishera territory is made up
of rather homogeneous, essentially terrigenous beds. Chochia (1955), who
traced this zone from west to east, distinguished three types of deposits:
western — composed of clays, sometimes carbonaceous shale interbedded
with dark limestone and thin intercalations of quartzose sandstone; central
and eastern — with predominance of deltaic deposits consisting of sandstone
and clays, less commonly with thin interlayers of limestone and carbonaceous

shales containing lenticular coal bodies; easternmost — consisting of inequigranular quartzose sandstones rich in plant remains with occasional
thin shale intercalations.

In the western type of section (Dzhezhim-Parma), limestone concretions
with remains of foraminifers and crinoids occur only in the uppermost part
of the coal-bearing suite.  On the strength of this, Chochia regards the entire section as being of shallow marine facies.

The section of the central and eastern parts of the territory, where the
sandstone content gradually increases eastward, has interlayers of clay,
shale and carbonaceous shale, as well as numerous lenses and lenticular
accumulations of coal.  The fauna in the occasional limestone interlayers
is exclusively foraminiferal.  Chochia classifies the area of development of
this type of the Yasnaya Polyana deposits as the zone of deltaic nearshore-
marine facies.

Chochia places the third type of section, 80—120 m thick and occurring
in the east of the Kolva-Vishera territory, in the continental facies zone.
Further east was located an erosion zone serving as the source of the terrigenous material.  Chochia writes that these conclusions are generally in
full agreement with the opinions held by Nalivkin (1943) and Vasil'ev (1950).
The latter authors maintain that the coal-bearing beds in the region stretching from the village of Kuzino in the south to the upper reaches of the Pechora River in the north are part of the deltaic formation of a single large
river which flowed from the Siberian continent and debouched into the sea
of the East European Platform in the vicinity of Kizel.  We have studied the
Kizel delta and proved that these viewpoints are erroneous, being based on
an insufficient amount of factual data on this almost inaccessible Kolva-
Vishera territory.

Somewhat better studied in this respect are the areas a little further to
the south and constituting part of the eastern continuation of the Kizel basin.
However, here too the marine sediments are truncated by a Recent erosion
surface.  Nevertheless, in the southeast of the basin, near the villages of
Druzhinino, Lysva and Kuzino, the coal-bearing arenaceous-argillaceous
rocks are gradually replaced by marine carbonates.  The influence of the
marine environment is also evident in the Skalny region, where siderite-
bearing sandstones contain remains of marine algae.  Even further south,
in the area of the settlements of Kyn and Obmanka, intercalations of limestone and dolomite are very common in the producing formation.

More to the south, along the foothill part of Bashkiria, marine carbonates
form a relatively narrow strip.  They are of Yasnaya Polyana age.  Vissarionova (1959) considers the following features to be typical of the section
of the Bobriki horizon in Ishimbai and Kinzebulatov: small thickness of the
terrigenous deposits which become increasingly calcareous.  Almost everywhere along the western slope of the Middle Urals the arenaceous-argillaceous deposits of this substage are replaced by massive limestones.  East
of Sterlitamak, in the basins of the Sikaza and Zigan rivers, the Middle Visean is composed of dark-gray and gray bedded limestones with abundant
remains of foraminifers and distinct species of brachiopods.  Nalivkin
(1948) compares these beds with the coal-bearing deposits of the Kizel and
Chus regions and the Moscow basin.

Coal-bearing beds have not been encountered on the western slope of the southern Urals; the Visean section there consists mainly of limestone. Only in the southernmost regions of the Mugodzhar Hills is the Yasnaya Polyana substage made up of arenaceous-argillaceous coal-bearing sediments.

All this suggests that during the Early and Middle Visean the region of the marine basin extended from the Pai-Khoi Range to the Mugodzhar Hills, and that a shallow sea with isolated islands of presumably volcanic origin occupied the site of the present-day Urals. The existence of the islands is corroborated by the petrography of the Lower Visean sandstones and conglomerates on the eastern slope of the Urals, the granitoids of which are of Caledonian age (data of Sergievskii and Petrenko (1949) and Smirnov and Smirnova (1953)). According to D. V. Nalivkin and V. D. Nalivkin, one of these islands in the Western Urals was located in the region of the Karatau Range. Smirnov (1957) believes that small islands existed both in this region and in the Hercynian intrusion zone. This conclusion is based on the fact that the Visean limestones unconformably overlie the Silurian effusives near the village of Karela, as well as on the presence of polymictic sandstone and conglomerate beds in the vicinity of the first Severny mine in the Middle Urals.

Continental deposits of the Bobriki horizon occur on the eastern slope of the Urals, indicating that the marine basin was relatively narrow and had the form of a peculiar strait linking two deeper seas. One of these seas must have been situated in the region of the present-day Barents and Kara seas, while the other was in the Caspian syneclise. In all probability, the width and depth of this strait were not uniform throughout.

Southeast of the city of Kungur, in the vicinity of Kuzino village, Smirnov (1957) noted that Tournaisian carbonate rocks are overlain by dolomitized limestone, calcareous shale and, in the upper part of the section, thin intercalations of quartz sandstone alternating with mudstones. Higher up in the section one encounters bedded clays with sandstone intercalations, containing Visean marine fauna. Somewhat further south, in the region of the Selikha River, the Tournaisian-Visean beds consist of persistent limestone.

The presence of a marine basin east of the Kizel region at that time is also borne out by the fact that to its south and north the terrigenous coal-bearing formations become thinner, mainly due to the decrease in thickness of the sandstone. The bedding of the latter also becomes less distinct, probably due to river-sea interaction.

In order to obtain a general picture of the Middle Visean Ural sea and its east coast, we shall briefly discuss the eastern regions of the Urals where a strip of terrigenous deposits, containing a number of coal deposits (Bredinskoye, Poltavskoye, Kamenskoye, Sukholozhskoye, Egorshinskoye, etc.), is developed. The coal-bearing beds are of a coastal-continental nature, but pass into marine carbonates toward the west. According to Smirnov, the limited occurrence of marine carbonates in the central zone of the Urals is due to the fact that their primary development was not continuous because of the presence of islands and the considerable erosion which followed. The western boundary of the Early Visean coastal portion of the land area has tentatively been drawn somewhat to the east of the Bredy and Kartaly villages; further north it passes near the towns of Troitsk, Kopeisk and Kamensk, and then continues to the mouth of the Pelym River and further north.

Lower Carboniferous (Middle Visean inclusive) sections have been studied in varying detail on both the eastern and western slopes of the Urals. North of the latitude of the town of Serov they are overlain by a thick cover of Mesozoic-Cenozoic rocks and are sometimes exposed on the western part of the slope. Arkhangel'skii (1953) has identified polymictic sandstones and conglomerates on the right bank of the Tynya River. The conglomerate pebbles consist of quartzite, chert, porphyry, limestone, quartz, arkosic sandstone and tuff.

In the south, deposits of the Visean stage are encountered in the Makhnevo, Alapayevsk, Yegorshino, Sukhoi Log, Kamensk-Uralski and Bogarak regions, and along the Miass River north of Chelyabinsk, as well as to its west, southwest and south. The sections of all these areas, which usually contain Visean coal-bearing beds, have been described by I. I. Gorskii, A. A. Petrenko, A. A. Pronin, E. A. Perepechina, G. A. Smirnov and others.

The most evident and distinctive feature of the coal-bearing beds on the eastern shore of the Early Visean Ural Sea is their polymictic composition. Also quite striking is the great difference in the thickness of the terrigenous formations on the west and east coasts of the paleosea. While the maximum thickness of the Lower Visean terrigenous beds is only 350—400 m in the regions of river deltas and in individual areas along the west coast of this basin or marine strait, it is several times greater on the east coast. For instance, the thickness of formations of this age in Bredy village is 1000 m; in the village of Poltavka, 900 m; in the vicinity of Borodinskoye village, 1200 m; along the Kabanka River, 1100 m; along the Miass River, 1000 m; near the town of Kamensk-Uralski, about 800 m; in the Sukhoi Log region, about 500 m; at Yegorshino, 1500 m (?); in the Makhnevo region, presumably about 1500 m.

The differences in the petrography and thickness of the terrigenous deposits are primarily indicative of the following: tectonic and physiographic conditions of formation; different source areas supplying material to the western and eastern parts of the marine basin; considerable differences in the elevation of the shores and in the steepness of the opposite continental slopes grading into the shelf.

According to Strakhov (1962), polymictic deposits like those along the eastern slope of the Urals are associated with an elevated relief and intensive mechanical denudation of the catchment areas. Rapid mechanical disintegration of the rock outcrops and fast mechanical removal of their fragments protect the latter from chemical weathering, even under conditions of a humid climate.

Olygomictic complexes were formed in a totally different physiographic environment. They accumulated in sedimentation zones of a humid plain where the relief of the drainage areas was leveled; the climate was not only humid, but also hot. The low relief of the territory reduced mechanical denudation and provided wide possibilities both for the chemical weathering of the rocks and their redeposition. This resulted in the total disintegration of the rock fragments and unstable heavy minerals, whereas the sand-silt fraction retained quartz and other stable minerals.

The easiest way to explain the difference in thickness of the terrigenous beds on the western and eastern shores of the marine paleobasin is to attribute it to different intensities of downwarping of the shelves. However, is this really the case? While recognizing the importance of the tectonic factor, we are inclined to agree with Strakhov that the actual mechanism responsible for the thickness of the humid formations is much more complicated and that the factors involved in this process are much more diverse. Strakhov believes that the rate of sedimentation in present-day basins is determined by four major factors: the ratio of the catchment area to the area of the terminal water body; the degree of dissection of the relief in the catchment area; the degree of humidity of the catchment area, i. e., the amount and regime of atmospheric precipitation; the propensity of the rocks in the catchment area to denudation and erosion.

While the dissection of the relief is controlled almost exclusively by tectonic conditions, the formation of sedimentary beds depends on a range of circumstances. One should therefore not attribute changes in the thickness distribution of this lithostratigraphic complex to tectonics alone, as is done by some researchers. Its formation is associated with many factors, significant among which are the climatic conditions and the action of rivers.

The character of the marine basin on the eastern margin of the East European Platform should be judged from the lithology of the sediments and the fossil fauna they contain. The limestones, both bedded and more massive, pure or interbedded with terrigenous formations, usually contain fossilized colonial corals, brachiopods and foraminifers, attesting to the normal salinity of the sea.

The mixed type of sediments and the presence of clastic varieties among carbonate rocks, as well as the thickness of the brachiopod shells suggest that the sea was shallow and possessed all features of an epicontinental basin.

### The Middle Transvolga Sea

The sediments that were deposited in the marine basin which once occupied the Caspian syneclise, in particular, in the southwestern part of the Kuibyshev or middle Transvolga region, have been studied in greater detail. The zone transitional from continental to marine environments of sedimentation has been traced into the Syzran region where Maksimova (1955) carried out an ecological-facies analysis of the producing formation of Yasnaya Polyana age. Here the producing formation consists of an alternation of siltstone, fine-grained sand, carbonaceous clay, limestone and sometimes thin intercalations of coal. The mentioned rocks occur repeatedly in the section, the thickness of individual layers ranging from 10—20 cm to 1.5— —2.5 m.

The sand and siltstone usually form elongated lenses of varying thickness. The limestone, which occurs as three or four layers in the terrigenous beds, is not persistent along the strike; it is fine-grained, argillaceous, slightly silty, with an admixture of fine carbonaceous mud and vegetal remains. In places it contains burrows. The sparse fauna is represented by brachiopods. Foraminifers and algae are found in the upper layers.

Very common in the producing formation are locally buried vegetal remains of woody species. This is confirmed by the presence of the bark of lepidodendron and stigmaria, as well as fragments of carbonized wood, imprints of large stems, and preserved stigmarian rhizoids. According to

Maksimova, stigmaria have been encountered in wells in the Syzran region, sometimes several per layer, and at different stratigraphic levels.

In addition to the plant material, there are remains of brachiopod and pelecypod shells, foraminifers and burrows of detritovores. The excellent preservation of the brachiopods indicates that they were buried in situ. Pelecypods are represented by only two species and are extremely rare. Crinoids, echinoids, corals and other groups of marine fauna characteristic of the Carboniferous are absent. These facts point to an environment quite different from that of a normal open sea.

The joint presence of wood remains and marine fauna in the basin is connected, in the opinion of Maksimova, with the development of low swampy islands and shoals on and around which there formed mangrove thickets, while the zones between the islands were characterized by sparse fauna. The ecological features of the latter suggest that the bottom sediments were soft unconsolidated muds.

After comparing the contents of chlorine in individual horizons of the producing formation, which vary from 0.17 to 0.90%, Maksimova arrived at the conclusion that the salinity of the basin during the period of deposition varied considerably, both in time and in area. The significant amounts of pyrite in rocks rich in organic matter show that the basin was at times contaminated by hydrogen sulfide.

Nearly all the features of the basin resemble those of the present-day zone of mangrove thickets on the Florida coast, which is also characterized by a complex alternation of land and sea areas, and by a nonuniform relief of the sea bottom. This environment is similar to the sedimentary environment of the Middle Visean producing formation in the Syzran region. The following are their common features: sediments of both continental and marine type; considerable variations in salinity and rapid changes of the bottom deposits over short distances; complex bottom relief, especially at the onset of deposition of the producing formation; paucity of the marine fauna. Mention should also be made of shoreline displacement and changes in the hydrodynamics of the basin, as these contributed to the further complication of the already rather chequered facies environment.

The width and extent of the "mangrove" zone have not yet been established. One thing, however, is clear. It passed into a more open though shallow marine basin. This is attested to, in particular, by the geochemical features of the depositional environments of the Yasnaya Polyana substage in the Transvolga region near Kuibyshev. Large-scale geochemical studies were conducted in this region in recent years, particular emphasis being placed on the lithology and mineralogy of the rocks, the equilibrium of the authigenic mineral forms of iron and sulfur, the amount and type of organic matter present, and the intensity of reduction processes.

The methods, procedures and results of these studies, aimed at clarification of the geochemical environment of sedimentation and secondary alteration of the rocks, as well as the geochemical interrelationships between the organic matter and the sediments, are discussed at length in the works of Radchenko and Fedorova (1949), Gulyaeva (1956), Rodionova and Konyakina (1957), Strakhov, Zalmanzon and Glagoleva (1959) and Vainbaum and Grigor'eva (1962). We shall only consider certain of the results obtained

by the above authors, since they may contribute to a more objective paleogeographic reconstruction of the marine basin which occupied the territory
of the Caspian syneclise.

On summarizing the results of the geochemical studies (including investigation of the overall content of organic matter, the hydrochemical regime of the sedimentation basin, the oxidizing-reducing environment, as well
as the processes of secondary reduction or oxidation of the rocks), L. A. Gulyaeva reached the conclusion that the sediments of the Bobriki horizon in
the southwestern part of the Transvolga region near Kuibyshev were deposited in a basin filled with brackish waters whose salinity was somewhat
lower than that of a normal sea, with a fairly high content of sulfates. The
mineralization of the basin waters was not constant, as evidenced by the
considerable variations in the chlorine content of the argillaceous rocks.

The regime of the basin southwest of the paleo-Kama delta zone, in the
region of the village of Pokrovka, was more stable and closer to marine.
In other regions (Bezenchuk, Zhigulevsk, Yakushkino, etc.) the water of the
basin was less saline. The rocks of the Bobriki horizon are characterized
by a rather high content of sulfide sulfur (1.68—4.15%), which points to the
intensity of the reduction process which was sustained, according to Gulyaeva, by sulfates which constantly diffused into the mud layer from the seawater.

In order to determine the extent to which the basin waters were contaminated with hydrogen sulfide, individual rock samples were treated with
dilute hydrochloric acid, and the content of ferrous and ferric iron in the
extract was then determined. Many of the samples either contained no
soluble iron or showed only traces of it. For instance, in the Bezenchuk
well the amount of iron not combined with sulfide sulfur does not exceed
2.5—5% of the total iron content. A similar low content of soluble iron in
conjunction with a considerable accumulation of sulfide sulfur (up to 4—7%)
was noted for the clays of the Bobriki horizon in other areas of the Transvolga region near Kuibyshev, thus suggesting a hydrogen sulfide oxidizing-
reducing environment of the marine basin. Many samples contain soluble
iron in variable amounts (from 0.1 to 0.3—0.4%).

As reported by Gulyaeva, 13 out of 31 samples of shale, mudstone, and
siltstone from the Bobriki horizon are virtually free of soluble iron, 7 samples
contain up to 0.15% $FeO + Fe_2O_3$, and only in 10 samples does its content
range from 0.2 to 0.4%. Rocks lacking soluble iron have been encountered
in each of the sections studied; they alternate with rocks containing up to
several tenths of a percent. The abundance of pyrite and absence of benthic
fauna indicate that an intensive process of sulfate reduction took place in
the muds of the basin. There were periods when large amounts of hydrogen
sulfide completely bound all the reactive iron, the excess hydrogen sulfide
finding its way into the benthic waters.

Consequently, Gulyaeva regards the rocks of the Bobriki horizon in some
areas of the Volga region near Kuibyshev as deposits of markedly reducing
and hydrogen sulfide types of oxidizing-reducing environments. The abundance of thin clayey sediments and the total absence of coarse clastic material in the sections are indicative of the tranquil nature of the basin, the
absence of strong currents, and relatively slow rate of sedimentation, which
contributed to relative enrichment in organic matter.

However, this conclusion applies only to that part of the Volga region near Kuibyshev which was occupied by a sea in Early and Middle Visean times. It is true that intensive formation of pyrite is characteristic of both the continental and the marine rocks of the Yasnaya Polyana substage. Yet, as pointed out by Strakhov et al. (1959), it is especially pronounced (percentage of total reactive iron) in the nearshore sand-silt sediments, usually dying out in more pelagic regions. The tendency of the authigenic iron to form pyrite and also the sulfurization of coal and large-scale pyritization are direct or indirect results of the enrichment of rocks with organic matter.

The marked increase in the content of pyritic iron in the continental deposits is attributed to the fact that the lake-swamp water bodies, which were situated on a lowland seacoast, were occasionally flooded by the sea, leading to an increase in their salinity and concentration.

The shift of pyritization toward the nearshore sand-silt zone of the marine basin is probably due to the high content of organic carbon in this zone. Strakhov considers the high mobility of the sulfides during the diagenesis of the muds to be an essential condition, as this enables them to migrate from sediment to sediment, in particular, from fine-grained pelitic sediments to coarser ones.

The content of organic carbon in these deposits is also fairly high. Even if the carbonaceous shales (rocks containing more than 10% organic carbon) are not taken into consideration, the average content of carbon reaches 1.78— —2.64% (Gulyaeva's data). The highest content of organic matter is encountered in clayey and, partly, silty varieties. The Bobriki horizon in the southwestern part of the Volga region near Kuibyshev does not contain humic acid, and consequently the organic matter in these deposits is derived from zoo- and phytoplankton.

The above considerations make it possible to state that the ancient sea in question was a shallow basin which became desalinated due to the inflow of river waters. It must have been the scene of vigorous development of both plant and animal life, as well as the large-scale accumulation of their remains. A reducing environment was the essential prerequisite for the formation of source beds from the nearshore-marine sediments of the Bobriki horizon.

### The Saratov-Volgograd Transvolga Sea

The marine basin in the Middle Volga region was not isolated; it linked up with the Ural Sea and extended far to the south, where we tentatively distinguish that part of it which was situated mainly in the Transvolga region near Saratov and Volgograd. In all probability, the physiographic environments of these seas did not differ markedly. However, they could have possessed specific features resulting from the effect of various river systems. While one of them was fed by the rather rapidly flowing paleo-Kama, which carried huge amounts of terrigenous material, the other received the waters of the more tardy Ryazan-Saratov paleoriver whose delta was somewhat smaller than that of the paleo-Kama.

The Bobriki deposits on the right bank of the Volga near Saratov are composed of the usual arenaceous-argillaceous, partly coal-bearing rocks which do not show uniform distribution. To cite an example, the deposits in the Yelshanka, Teplovka, Kazanlin and other northern areas occur only as isolated patches confined to narrow erosion enclaves. The zone of increased thickness of the Yasnaya Polyana substage can readily be traced into the southern regions (Peskovatka, Goryuchka, Ilovlya, etc.). In the Sovetski, Stepnoi and other areas of the Transvolga region the thickness of this zone gradually decreases. On the right bank of the Volga the rocks of the coal-bearing horizon are predominantly littoral-continental; in the Transvolga region the terrigenous deposits contain numerous interlayers or lenses of limestone, dolomite and siderite typical of a nearshore-marine depositional environment.

The section of the Tula horizon in the Transvolga region is made up largely of alternating clays and limestones which are present in approximately equal proportions. The lower half of the Tula beds in the regions of Bogayevka, Surovka, Vyazovka, Peschany Umet is notably arenaceous, while its upper part shows an increase in the amount of carbonate rocks. Further south, in the Goryuchka, Peskovatka and Ilovlya areas, the role of carbonates in the Tula horizon becomes even greater. The sandstones pinch out almost completely in the Transvolga region, and only in the sections of the Generalski and Sovetski regions does the clay still contain thin intercalations of siltstone.

The lithology of the Middle Visean beds on the left bank of the Volga near Saratov points to the former existence there of the coastal zone of a shallow sea. This is corroborated by the repeated alternation of terrigenous, mainly clayey sediments and carbonates, the presence in the section of numerous inclusions of finely disseminated pyrite, in places forming large or continuous interlayers, and the convoluted burrows of detritovores in the sand-silt rocks. Fairly widespread along the ancient marine shore are peculiar strips of alluvial and beach sand deposits.

According to Kotova, Latskova et al. (1958), the terrigenous sediments of the Yasnaya Polyana substage in the Volga region near Saratov contain, on the average, about 2% organic carbon, whereas the carbonate rocks contain only 0.5%.

The average content of sulfide sulfur in the terrigenous deposits of this substage is 1—2%, while in the rest of the section of the Visean stage it amounts to mere fractions of a percent. This suggests the reducing or even strongly reducing environment of deposition of the producing formation. This is also supported by the predominance of soluble ferrous iron over ferric iron.

The sediments of the upper part of the Tula horizon, which is made up of thick limestones containing fossilized remains of brachiopods, foraminifers and corals, were deposited in a more open sea with normal salinity. Characteristic for the Tula limestones are colonial and solitary tetracorals, as well as bryozoans. The most common brachiopods are productoids.

Consequently, during the Middle Visean, a considerable part of the territory of the Transvolga region near Saratov was occupied by a shallow, somewhat desalinated marine basin with environments quite favorable for the accumulation and burial of organic remains which could serve as the source material for the formation of oil. Similar conditions are known to have existed both to the northeast (Transvolga region near Kuibyshev) and south-west (southeastern Volga region near Volgograd). In other words, a single marine basin occupied the Caspian syneclise and received the waters of two large, closely spaced paleorivers.

The coast of the marine basin, alternately taking the form of swamped dry land areas and wave-built sand beaches and bars, must have changed its appearance repeatedly throughout Yasnaya Polyana times. The relatively diffuse network of boreholes makes it almost impossible to fix the position of the shoreline of this ancient sea. Nevertheless, the arbitrary boundary between the dry land and sea drawn on the paleogeographic map is probably a fair approximation of the position of the ancient shoreline.

There is no need to cite the geochemical characteristics of the organic matter and bitumens occurring in the beds under study in the lower reaches of the Volga, since they differ but little from those described above. We shall only present some conclusions from geochemical studies of the bitumen and oil in the Volgograd region (data of Tverdova, Astashova et al. (1962)).

1. The organic matter is totally independent of the distribution of bitumen and oil.

2. The background content (0.02) of the bitumen prevails throughout the section.

3. Higher contents of bitumen were recorded in the oil- and gas-impregnated beds within the oil- and gas-producing region.

4. The character and type of the bitumen depend on the lithology of the containing rocks. For instance, bitumen A derived from argillaceous rocks is aromatized and oxidized, whereas the same bitumen isolated from sandstone and limestone contains a larger amount of paraffin structures of high molecular weight. The oil content of bitumen from argillaceous rocks does not exceed 30%, and the ratio of benzene to alcohol-benzene tars ranges from 0.21 to 1. The oil content of bitumen from arenaceous rocks attains as much as 65%, and the ratio of benzene to alcohol-benzene tars is about 2, thus indicating the obviously petroleum nature of the bitumen. The content of oil in the carbonate rocks is intermediate (50%).

5. The type of bitumen is independent of the geological age of the containing rocks.

6. All the bitumen studied, irrespective of the lithology of the containing rocks, is secondary epigenetic bitumen.

The above authors believe that all types of petroleum in the Volga region near Volgograd are genetically similar, the Bobriki horizon having the highest oil content.

The above data also confirm the former existence of conditions which promoted intensive bitumen formation in the Lower Carbonaceous producing formations. Such conditions primarily existed in the nearshore-marine zone of the Middle Visean sea, which we are inclined to include in the region of oil and gas formation.

The sea in this region at times penetrated as far along the ancient river valley as the region of Umet, where limestone is interbedded with the sandstone and clay of the Bobriki horizon. Further to the southwest the shore of the marine basin extended somewhat to the north of Volgograd, and then continued northwest, skirting the southern slope of the Voronezh anteclise. This follows, in particular, from the section of the Middle Visean deposits in the Marinovka area, situated west of Volgograd, near the village of Kalach on the Don. The lithology of the Bobriki horizon here differs rather markedly from that of formations of the same age on the right bank of the Volga near Volgograd.

The Bobriki horizon is made up of dark, compact, slightly micaceous clay (somewhat silty, calcareous, in places strongly pyritized) with intercalations of dark gray calcarenite and fine-grained argillaceous limestone (compact, recrystallized, with small clay gouges on the bedding planes). The clay contains numerous concretions and lenses of siderite, as well as many pyrite concentrations. The Tula horizon is represented by homogeneous limestones, argillaceous to varying degrees, among which occur inequigranular, fine-grained, and relict-organogenic varieties. The latter consist of 30—40% detritus of small foraminifers and brachiopods and 60—70% cement. Locally, the limestone shows transition to marl. Throughout the section of the Tula horizon freshly fractured rocks give off a strong bituminous odor.

It is quite evident that during the Middle Visean a shallow sea with limans, bays and lagoons existed in southern parts of the Volgograd region. The conditions here were more stagnant-reducing than those in the open parts of the basin. During the Late Tula the sea became somewhat deeper as its boundaries expanded.

Data from deep wells that penetrated the downwarped portion of the southern slope of the Voronezh Massif between Millerovo and Glubokaya (near the villages of Tarasovka and Novoaidar, north of Lugansk) reveal that the Middle Visean sediments directly overlie crystalline rocks of the Precambrian basement and have a nearshore-marine facies (alternation of clay, sandstone and limestone).

### The Dnieper-Donets Sea

The Elenovka sedimentary complex of the Donbas, to the northwest of the southern regions of the Volga near Volgograd, in the southeastern part of the Dnieper-Donets basin, corresponds to the Yasnaya Polyana substage (D. E. Aizenverg). In the Donbas this interval is composed mainly of homogeneous fine-grained organogenic limestone. The lower part of the Elenovka complex contains several thin intercalations of calcareous shale interbedded with flaggy limestone. In the upper part there are peculiar thinly bedded rocks which are termed siliceous marl.

The mineral composition of the limestones is characterized by a predominance of light minerals — quartz and secondary silica [sic].* The principal heavy minerals are hydrogoethite, ilmenite, magnetite, pyrite and zircon.

The sediments of this complex contain a considerable amount of organic remains, the limestones being either foraminiferal or coral-foraminiferal. The siliceous marls in the upper part of the complex consist of dark, thinly bedded argillaceous rocks with a finely elutriated homogeneous groundmass which has a strong bituminous odor and readily effervesces when treated with HCl. Typical siliceous marls are rich in fauna (sponge spicules, radiolaria and brachiopods).

* [By definition, a limestone consists chiefly of calcium carbonate (calcite) which is regarded as a light mineral.]

The lithology of the carbonate beds and the character of the numerous faunal remains buried in them suggest that the Donets sea differed from the marine basin that fringed the eastern margin of the East European Platform. Apparently, the Dnieper-Donets sea had the form of a gulf which penetrated far inland between the Voronezh and Ukrainian shields. It is quite possible that its shelf zone fringing the shields was also relatively steep and narrow. Judging from the sediments which are exposed in the vicinity of the Kalmius River in the southwestern part of Donbas, they were deposited in a tranquil marine environment, far from the coastal zone. This indicates that during the Middle Visean the nearby southeastern closure of the Ukrainian Crystalline Massif, known as the Azov block, was covered by a sea whose sediments were later eroded.

According to Aizenverg, the paleontological remains and, especially, the considerable proportion of endemic faunal elements are indicative of limited links with other marine basins during a certain period of the Early Visean. However, the composition of the fauna in the upper part of the carbonate beds of the Elenovka complex (Yasnaya Polyana age) differs appreciably, reflecting the subsequent broader links with the adjacent basins. The Late Tula sediments show a somewhat different facies indicative of the fact that the open sea became shallower while retaining its normal salinity. The character of the fauna supports the assumption that this sea had free connections with other parts of the marine basin, which became deeper to the southeast while in the northwest it was gradually filled with a thick series of terrigenous sediments.

The marine deposits of the Elenovka complex are distributed along the northern slope of the Ukrainian Massif. At the confluence of the Solenaya and Volchya rivers they appear as rather homogeneous beds of fissured argillaceous limestone alternating with shale. Further west, in the Pavlograd region, the section contains additional clayey intercalations but the overall thickness of the complex decreases. The rocks here locally contain abundant unidentified faunal remains, commonly pyritized.

Even further to the northwest, in the Novomoskovsk area, the lower part of the Elenovka complex incorporates a small (3—6 m) arenaceous-argillaceous member with a few limestone intercalations. Argillaceous-carbonaceous beds occur in the uppermost part of this member. The upper part of the complex consists of argillaceous limestone interbedded with gray shale. In a few places the rocks at the top are silicified, which may indicate a gap in sedimentation.

Consequently, as the series thins out to the northwest, the carbonate content decreases while the role of terrigenous deposits in the section increases. Although the limestone interlayers contain abundant foraminiferal fauna, suggesting a normal marine sedimentary environment, poorly preserved remains of other animal groups are also encountered in this area, indicating an unstable marine regime, probably due to the proximity of the shoreline and the influence of rivers flowing into the gulf.

In the northwestern part of the Dnieper-Donets basin the gulf became even narrower; its shore was located approximately between Poltava and Chernigov, the site of the delta of an assumed paleoriver.

Even in the lower reaches of the Orel River the Elenovka complex starts with a terrigenous member consisting of sandy shale and inequigranular quartz sandstone often rich in carbonized plant remains. The member contains occasional argillaceous intercalations with ostracod fauna; in the top part of this member there is a limestone interlayer with foraminiferal remains. Still higher up one encounters shale or sandy shale with intercalations of argillaceous limestone and occasional sandstone members. These are overlain by argillaceous, sometimes sandy limestone (containing foraminifers and ostracods) interbedded with shale. The limestone in the uppermost portion of the complex contains numerous sponge spicules and impoverished foraminiferal fauna.

Northwest of Poltava, in the Sagaidak, Radchenkovskaya and other areas covered by oil exploration, the carbonate rocks progressively give way to pyritized calcareous-argillaceous rocks, in places with inclusions of siderite, sometimes with sandstone intercalations which are not persistent along the strike.

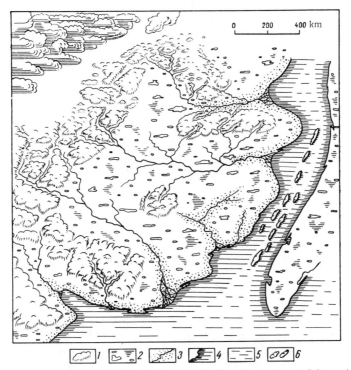

FIGURE 57. Paleogeographic map of the East European Platform during Yasnaya Polyana times:

1 — elevated peneplaned relief forms; 2 — lacustrine-palustrine plain; 3 — rivers and their valleys; 4 — nearshore-marine shoals; 5 — deepwater area; 6 — islands.

A petrographic study of the Middle Visean limestones in the Radchenkovskaya area shows that their light minerals consist of secondary silica represented by secondary quartz in the form of sponge spicules or unidentifiable platy fragments, less frequently by grains showing regular crystallographic outlines [sic].* The heavy minerals consist essentially (95—99%) of pyrite and minor amounts of zircon.  As one approaches the margin of the north-western part of the basin, which generally corresponds to the coastal zone of the Middle Visean gulf, the role of carbonates decreases, and the section is dominated by terrigenous rocks represented by argillaceous and inequi-granular arenaceous varieties.  According to data from three wells, the rocks in the Radchenkovskaya area have the following composition (on average): 24.4% sandstone and siltstone, 29.1% argillaceous rocks, and 46.5% limestone.

In the Romny-Zasulye region the Elenovka complex is represented large-ly by terrigenous deposits with minor intercalations of limestone.  Sand-stone members (quartz or quartz-mica, medium- and coarse-grained, sometimes with quartz pebbles, locally calcareous) are more frequently en-countered.  All this suggests not only that the sea was shallow on the given territory, but also the proximity of its shores.

The abundance of siderite formations in the deposits under study and their partial pyritization are indicative of an oxidizing-reducing environ-ment in this part of the marine basin.  However, the abundance of foramini-fers and in some cases the presence of other organic remains indicate slight contamination of the benthic waters by hydrogen sulfide.


Now that we have specified the areas of distribution and the character of the seas and land areas, as well as the major hydrographic network, of the East European Platform during Yasnaya Polyana times, it is possible to en-visage the paleogeographic environments of sedimentation (Figure 57).  Al-though one is entitled to question certain details on the schematic map, its main significance lies in the fact that it provides a fundamental interpre-tation of a vast body of factual evidence.  The environment of the remote past shown on the map determined the present distribution of fossil fuels within the substage under study, as will be shown later on in the book.

### Paleoclimate

Many researchers working in the field of climatology have established that climate depends on various factors which are tentatively divided into two main groups: astronomic and geologic.  The first group includes changes in the amount of solar energy with time and its varying supply to the earth; the second group incorporates the composition of the atmosphere, the distri-bution of land and sea, polar wandering, the elevation of the continents, orog-raphy and hydrography, and other geologic factors.  It is sometimes diffi-cult or even totally impossible to draw a line between these two groups. One thing is clear: the climate changed repeatedly and undoubtedly had a marked influence on sedimentation and the development of organic life.

Strakhov (1945) is of the opinion that climate played a major, decisive role in determining the lithology and geochemical character of sediments.

---

* [See footnote on p. 179.]

In his words: "While oscillatory movements of the earth's crust determine the very possibility of sedimentation and its geographic localization, the lithology of the sediment and its geochemical character, being dependent on the course of the chemogenic sedimentation, are primarily determined by the climate of the region where the tectonic factor 'makes sedimentation at all possible'."

Consequently, the lithofacies features of the Middle Visean producing formation may be considered as an indicator of the climatic environment. The climatic conditions are particularly well recorded by deposits of the coal-bearing formation and especially by the coal beds themselves. The latter originate during swamping which generally depends on the climate. Peat can form only under conditions of excess humidity.

Numerous remains of Lower Carboniferous thermophilic and hydrophilic plants are found on the East European Platform. These include lepidophytes, calamites, ferns and other types of woody vegetation of the so-called West-phalian zone. The Lower Carboniferous is characterized, in particular, by extensive development of a rich flora of the higher sporophyte pteridosperms.

We have repeatedly emphasized the humid nature of the Middle Visean land area, which was primarily due to the humid climate. Abundant precipitation enhanced erosion processes and the rivers supplied vast amounts of detrital material to the marine basins. This is borne out by the typical features of the mainly terrigenous sedimentation in the epicontinental seas considered above. Humid climatic conditions are also indicated by the wide-spread occurrence of tree ferns in the Lower Carboniferous. These plants require for their development not only high temperatures but also abundant humidity.

The presence in individual areas of the East European Platform of rather distinct weathering crusts with kaolinite and bauxite deposits, the extensive coal-bearing areas, and the numerous traces of vigorous river action suggest that the climate during the Middle Visean was humid. It is much more difficult to reconstruct the temperature regime of the ancient climate. However, the available samples of carbonate rocks from the Dnieper-Donets and other seas, the relative abundance of foraminifers and the numerous finds of corals and other thermophilic fauna point to the fact that the marine basins were located in the subtropical or even tropical zone.

Moreover, according to A. N. Krishtofovich, the morphological features of Carboniferous plants suggest a warm and humid climate, since several of their features coincide with those of present-day tropical vegetation. The absence of annual rings in Carboniferous wood shows that its growth continued uniformly throughout the year and was uninterrupted by either cold or dry seasons. It follows from this that the belt of luxuriant vegetation, which furnished the source material for coal accumulation, was situated in a climatic environment similar to that of the present-day tropics or even the equatorial zone.

It is of interest that the Lower Carboniferous woody vegetation of the Spitzbergen Archipelago is also of the subtropical type. According to the reckoning of A. N. Orvin (1940), the average annual temperature in the archipelago was at that time 30° higher than at present.

Chemogenic limestone and, especially, primary dolomite are known to have been formed in zones of warm or hot climate. Both types of rock are very common in nearshore-marine deposits of Yasnaya Polyana age on the East European Platform.   The tropical nature of the Lower Carboniferous fauna has been pointed out in several works by M. D. Zalesskii, A. N. Krishtoforovich, G. Potonier and other researchers.

It is very difficult to reconstruct the role of the wind as an element of the paleoclimate, although it is quite evident that cyclones and other atmospheric disturbances caused frontal precipitation on a huge scale.   The hurricane winds of tropical and subtropical regions brought heavy showers which undoubtedly affected the hydrodynamic regime of the rivers.   Moreover, the wind played an important role in the transport of fine-grained detrital material, volcanic ash and dust.   The predominant wind direction is usually reconstructed from the shape of ancient dunes, wind ripples, etc. Such relevant data are not available for the territory in question.   However, Schwarzbach (1955) has reported that the roots of fossilized trees in Carboniferous deposits near Sheffield extend horizontally on the westward side, while on the side facing east they penetrate deeper.   This is exactly how the roots of trees exposed to westerly winds are anchored in the ground at present.   This led the mentioned investigator to assume that the predominant winds in Europe during the Carboniferous Period had approximately the same direction.

It is quite probable that the Middle Visean climate was characterized not only by heavy rainfall, but also by frequent thunderstorms which could have caused forest fires.   The finds of charcoal among alluvial sandstones of the paleo-Kama are not incidental.   It is perhaps of interest to give the present figures for thunderstorm days recorded annually: 322 in Java; 100 in India; and only 17 in Europe.

In our studies we have given particular attention to the aspect of mechanical transport of weathering products by river waters, which is connected with the climate.

The geographic distribution of the degree of mechanical denudation is a most characteristic phenomenon, as noted by Strakhov.   He distinguishes two parallel latitudinal zones with markedly differing features.   The first is the fairly humid climatic zone of the Northern Hemisphere; the annual isotherm of +10°C is its southern boundary.   Annual precipitation in this zone amounts to approximately 150—600 mm.   The overall intensity of mechanical erosion here is either low or moderate; in most cases it is of the order of less than 10 tons/km$^2$.

The second zone includes the subtropical and tropical humid belts.   It lies between the +10°C isotherm in the Northern Hemisphere and the +10°C isotherm in the Southern Hemisphere.   Over most of its territory the average annual temperature does not fall below +20°C.   Annual precipitation amounts to 1200—1300 mm.   The intensity of mechanical erosion is much higher than that in the northern temperate zone and is generally of the order of 50——100 tons/km$^2$, in some areas rising to or even exceeding 100—240 tons/km$^2$. For instance, mechanical erosion in  Southeast Asia averages 390 tons/km$^2$, while in the basins of the Indus, Ganges and Brahmaputra it exceeds 1000 tons/km$^2$.

Strakhov arrived at the conclusion that mechanical erosion on a territory with a humid climate shows distinct zonality.   Zones with fairly humid climates undergo weak to moderate mechanical erosion, while tropical and subtropical zones experience strong to very strong denudation. Strakhov maintains that "together with the relief, the climate, in particular the temperature regime and the amount and conditions of atmospheric precipitation, plays an important role. It is natural that as the tectonic activity of a region is heightened, this is followed by an increase in the energy of mechanical erosion."

The regular effect of climatic zoning on the ongoing process of denudation provides us with a clearer picture of the conditions under which detrital material was mechanically transported by Middle Visean paleorivers and enables us to assess the contribution of climate to the complex process which gave rise to the producing formations under study.

To this day, many geologists still hold the view that the tectonic regime is the sole factor controlling the thickness of sedimentary formations. Strakhov, however, has convincingly proved that the mechanism actually responsible for the thickness of such rock formations is much more complicated and that the factors involved are much more diverse. According to Strakhov, the average intensity of sedimentation in present-day basins is determined by the following formula:

$$m = \frac{B}{L} RHD,$$

where $\frac{B}{L}$ is the ratio of the catchment area $B$ to the area of the terminal water body $L$; $R$ is the degree of dissection of the relief in the catchment area; $H$ is the degree of humidity of the catchment area or in other words the annual amount and regime of precipitation; $D$ is the erodibility of the rocks in the catchment area. The average intensity of annual sedimentation is proportional to each of the above values, i.e., factors.

Strakhov admits that the above formula is not readily applicable to the analysis of ancient deposits, since it does not take into account the intensity of tectonic movements within the sedimentation region. Nevertheless, the dissection of the relief is both directly controlled and preserved for a long period of time by the intensity of the tectonic regime; for that reason the value $R$ in the formula is replaced by the value $T$ which expresses the intensity of tectonic movements, assuming that $R = aT$ where $a$ is a certain conversion factor. The formula then becomes:

$$m = \frac{B}{L} (aT) HD.$$

Strakhov claims that studies of ancient sedimentary formations fail to take account of the influence of humidity ($H$) and the erodibility ($D$) of the rocks in the catchment area, although they must also have played a significant role in the past. This follows from the known fact that at all times the amount of atmospheric precipitation in humid tropical zones greatly exceeded that in humid temperate zones; monsoon climates, with their characteristic alternation of periods of drought and rainfall, have always existed; the earth's surface has always been made up of rocks that differed in their erodibility. Hence, the mechanism governing the thickness of ancient sedimentary series must, in principle, have been similar to that functioning at present, i. e., it was the sum of numerous factors among which "the tectonic factor must have been the major, although by no means the only one."

It appears that the effect of climate or the humidity factor on the thickness distribution of a sedimentary formation can be determined by comparing the rates of sedimentation in different parts of the ancient landscape.

The much greater thickness of the terrigenous beds of the Yasnaya Polyana substage in river valleys and deltas than on the adjacent territories is primarily due to the hydrodynamic regime of the rivers, which was particularly dependent on the climatic factor, rather than to the tectonics. The rivers had a pronounced effect on the rate of sedimentation in shallow seas and, in particular, in estuaries.

We concur with Strakhov that the average rate of sedimentation in basins of the geologic past and the range of its variation are generally in good accord with present-day "norms," and therefore we offer the following comparison.

The present rate of sedimentation in the northern part of the Caspian Sea is 100 cm of wet or 30 cm of dry sediment per 1000 years. If one assumes roughly the same rate of sedimentation in estuarine areas of the Middle Visean seas, then the buildup of the terrigenous nearshore-marine beds, 250—300 m thick, must have taken 1 million years. However, this must have taken place in areas where enormous amounts of mechanical erosion products were supplied by rivers (similar to the present-day Volga) from the catchment areas to the marine basin. According to Ronov (1949), the average rates of sedimentation per 1000 years in the Lower Carboniferous water bodies of the East European Platform amounted to 2.2 cm in the Volga-Ural region, 0.9 cm in the Moscow syneclise, 0.5 cm in the Dnieper-Donets basin, and 7.0 cm in the Ural geosyncline.

Red rocks of Yasnaya Polyana age, occurring on the southeastern slope of the Baltic Shield and partly in the Timan-Pechora province, may provide us with some idea of the paleoclimate. They suggest that these regions experienced an alternation of rainy and semidry seasons; even during the semidry seasons individual moist areas, either near lakes or along river courses, remained covered by vegetation. Rivers also played an essential role in the deposition of the red beds. During rainy periods the rivers flooded huge areas and deposited sediments. During the dry seasons the rivers became shallow or possibly even dried up; weathering and intensive surface oxidation of the clay deposits produced the red-brown coloration of the soils and the weathering crust.

Rukhin (1959) reported that the most widespread facies in the red beds is the alluvial floodplain facies. This facies probably includes individual layers of the producing formation in the Timan-Pechora region. Some silt-clay and clay-carbonate sediments deposited on the bottom of lakes, oxbows and limans obviously consist of redeposited red rocks. Admixtures of organic matter in some places changed their coloring to bluish-gray, as can be observed in sections of the middle reaches of the Pechora basin, where sticky variegated clays (brown-mauve, bluish-green, claret, and red) are very conspicuous.

Red beds are sometimes deposited on alluvial plains adjacent to lagoonal zones and shores of shallow seas. The courses of some rivers on such plains are very unstable, also primarily due to features of the climate.  For instance, the Hwang Ho has caused more than 1500 disastrous floods during the last 4000 years, and has changed its course more than 20 times, wandering over a territory equal to that of France.

The Middle Visean Timan-Pechora paleoriver, whose deltaic region alone was more than 400 km wide, could have undergone similar changes of course.

In order to obtain a clearer picture of the paleoclimate, one has to consider the territory being studied in relation to geographic paleolatitudes. It is common knowledge that climatic zoning has not been a constant feature throughout the earth's history.  It is due to changes in the position of the earth's axis of rotation and, as a consequence, the shifting of the poles, as is borne out by paleomagnetic observations.  The latter are based on the fact that certain sedimentary rocks are capable of retaining remanent magnetization.  The ferromagnetic minerals incorporated in the sediments either at the time of their deposition or prior to their lithification were arranged in such a way that their magnetic axes coincide with the direction of a local geomagnetic field that existed at the time of sedimentation.

This natural remanent magnetization is often preserved to this day and enables us to reconstruct the magnetic field of the earth in former times.

Paleomagnetic and other data appearing in the works of Strakhov, Krapotkin (1958), Khramov (1958), Rukhin (1959) and others point to the fairly considerable wandering of the earth's poles, starting from the Proterozoic. According to certain data, the magnetic north pole during the Lower Carboniferous was located in the Far East of the Soviet Union, while according to others it was in the Pacific Ocean.  The paleomagnetic observations are in accord with paleobotanic, paleofaunal and geochemical data, which also suggest that during the Middle Visean the tropical zone extended at a sharp angle to the present-day lines of latitude and had an almost meridional orientation relative to the geographic coordinates of today.

Thus, the basic zoning of the tropical and subtropical humid landscapes on the East European Platform during this time interval has a general northwesterly orientation.

## Occurrence of Coal as an Indication of Physiographic Environments of Sedimentation

Oil- and gas-bearing formations fairly often contain deposits, beds, lenses or intercalations of coal.  This phenomenon of joint occurrence has been observed in the Appalachian basin in the U. S. A., in the Volga-Ural region, in West Siberia, as well as in many other regions around the globe.  The accumulation of coal is a distinct manifestation of a certain physiographic environment which is in some cases of interest to petroleum geologists. To date, insufficient attention has been paid to the presence of coal as a source of information on the physiographic environment.  Therefore, we find it timely to discuss this aspect in greater detail.

Accumulation of coal starts with the formation of peat — a biochemical process taking place in inland water bodies (shallow lakes, marshes, quagmires, and other types of swamps) saturated with nutrients for the development of a plant community.  Abundant moisture, i. e., the hydrological regime controlled by the climate and local geomorphology, is the main factor governing peat accumulation. The filling of lakes and swamps with peat is especially rapid under warm and humid climatic conditions.

Gravitational forces are sufficient for the gradual accumulation and settling of the peat layer to the bottom of the water body. However, the stronger the general subsidence of the sedimentary basin, the more rapid the changes in the conditions of peat accumulation, which sometimes even ceases. The water bodies became shallower as clastic material was supplied from the shores and organic ooze accumulated.

The accumulation of peat was particularly irregular in river valleys and deltas. Notwithstanding the significance of tectogenesis, the role of the paleogeographic environment in coal accumulation remains decisive.

How can this environment be reconstructed? One can arrive at the answer by considering the experience gained over the years from studies of the characteristic features of coal occurrences and their mineral composition in the Volga-Ural and other regions. We have based our studies on material from oil exploration drilling. Below we shall discuss a few regions where oil and gas occur in deposits which also contain coal.

### Middle Reaches of the Volga

In the southeastern part of Tataria, in the Suleyevo-Tashla area, well 41 penetrated a 15-m-thick coal lens at a depth interval of 1053—1068 m. In well 611 in the same area, 5 km to the northeast, the coal is one meter thick. Eighteen kilometers east of well 41, well 21 drilled in the Aznakayevo area penetrated a 7-m-thick coal bed at a depth interval of 1062—1073 m.

A rather thick (8 m) coal lens was encountered in well 33 in the extreme east of Tataria, in the Muslyumovo region. Individual wells in the north of Tataria (Naberezhnye Chelny and Yelabuga regions), in the south (Cheremshan region) and in the west (Ulema area) have revealed the presence of coal deposits. Coal beds and interlayers of varying thickness were also discovered in the Aktash, Sarailin, Buldyr, Melekess, Kirmen, Kazaklar and other areas. The enclosing rocks here consist of the usual terrigenous sediments: dark to black mudstone and carbonaceous shale, and gray and dark gray, often argillaceous siltstone and sandstone which are interbedded and display frequent changes in composition and thickness.

The coal-bearing layer penetrated by the Suleyevo 41 well is 40 m thick. It is underlain by Tournaisian limestones at the top of which there is light-gray quartzose sandstone. Higher up in the section there occurs a 1.5-m-thick bed of dull, compact coal. It is overlain, in turn, by a 20-m-thick member composed of sand-silt rocks with mudstones at its top and base. This member contains an intercalation of dull black coal, 0.3 m thick.

Still higher up there is a thick coal bed divided into three parts by thin mudstone intercalations. The lowermost part of this bed, or, more specifically, the lenticular deposit, is made up of bituminous, semidull, in places semibright or even bright varieties of hard, compact, sometimes laminated coal (8.8 m thick). This part of the coal deposit is separated from the middle one by a thin (0.05 m) mudstone intercalation which is overlain by semibright and bright coal (0.95 m). Another intercalation (0.1 m) of carbonaceous mudstone is followed by the uppermost part of the coal deposit, consisting of a relatively uniform mass of dull and semidull, compact, in places laminated coal (4.85 m). The total thickness of the three parts which virtually constitute a single coal body is about 15 m according to the core samples, and about 17 m according to the logging data. The structure of the coal deposit is indicative of the fact that during its formation there were periods when peat accumulation was interrupted.

The coal-bearing beds are topped by a layer of dark-gray laminated mudstone (2.8 m) with thin intercalations of siltstone. Higher up the well penetrated gray fine-grained oil-bearing sandstone forming the base of the Tula horizon.

A rather detailed petrographic study of core samples of coal from the Suleyevo 41 well was performed by A. P. Bludorov who found that the predominant coal microcomponent is spore durain, occasionally developing into spore clarain. The coal consists of macerals and phyterals. The latter are represented by pieces of stems, spore coats and cuticles.

Vitrain occurs as strips 0.2—0.5 mm wide, less commonly 1.0 mm. These strips pinch out in some places.

Xylovitrain is present in the coal in small amounts and appears in thin sections as small brown fragments. Sometimes xylovitrain grades into an amorphous mass.

Fusain is very common and appears in thin sections as small lenses or strips extending across the slide.

Fairly large amounts of macro- and microspores are visible in all thin sections.

The coal from well 41 is characterized by the following indexes (in %): 3.6 moisture; 14.6 ash; 44.0 volatiles; 4.3 total sulfur; 77.42 carbon; 4.1 hydrogen; calorific value 7400 cal; density of damp coal 1.26 g/cm$^3$, density of dry coal 1.18 g/cm$^3$. These physicochemical properties of the coal from the Suleyevo area indicate that it is of the hard, long-flame variety. In individual cases the degree of coalification approaches that of the gas coal in the Gremyachinsk field in the Kizel basin.

The petrographic composition of the coal being studied is indicative of the conditions under which it was formed. For instance, the presence of vitrain in thin sections attests to the fact that the water level varied during the accumulation of the vegetal matter. Fusain, which is also a common constituent, is produced as the result of decomposition of the plant material with participation of atmospheric oxygen. The starting material could have been small fragments of partly decayed wood remains. Consequently, the presence of fusain and fusain-xylain indicates a low water supply to the peat bog. This is also attested to by the predominance in the Suleyevo well of dull coal which is usually produced under subaerial conditions, i. e., with a very thin water cover. However, among the dull coal one also encounters semi-bright varieties, the formation of which calls for anaerobic conditions with more abundant water supply.

The most common petrographic type of coal, identified according to the ratio of phyterals to macerals, is spore durain; intermediate varieties are less frequently encountered.

Spore durain is either a transparent or opaque mass with a predominance of phyterals. The latter are represented largely by macro- and microspore coats, sometimes also constituting the macerals. Thin sections also show fragments of cuticles, sometimes also clay gouges and admixtures of quartz grains and pyrite.

The following durain varieties are present: woody and spore-woody durain, and also clarain-durain. The woody durain consists of reddish-brown or black macerals, and subordinate amounts of phyterals including spore exines and fragments of xylovitrain and fusain, the latter predominating.

The phyterals of spore-woody durain are made up of approximately equal amounts of macro- and microspore coats and fusain. The clarain-durain, like the preceding types, consists of macerals and phyterals, the concentration of the latter varying.

Under the microscope remains of micro- and macrospores and also cuticles are distinctly visible in thin sections. The microspores, which according to the shape of their coats are of the shagreen type, are sometimes present in amounts not only exceeding all other phyterals but also the macerals. Such a coal structure is characteristic of liptobioliths, which are produced as a result of total decay of the less stable lignite-cellulose tissues and the accumulation of more stable vegetal matter. Microspore coats (exines), compressed parallel to the stratification, are usually observed in thin sections.

Considerable amounts of macrospores are also visible in nearly all thin sections. They are usually yellow (less commonly orange-colored) and are typical of the phyterals in the coals in question. The macrospores usually appear as smooth or tubercular exines showing various degrees of preservation.

The cuticles observed in the thin sections have the form of thin yellow strips extending parallel to the bedding; sometimes they are dotted on one side with small serrated tubercles.

The above-described coal types are characteristic of most of the territory of Tataria and some other regions of the Middle Volga region. All of them were formed under continental conditions from remains of terrestrial plants which accumulated in water bodies within the paleorelief of a lowland plain. These are almost exclusively humic coals, suggesting that the plain was covered with swamps and small lakes overgrown with vegetation, where sapropel coals could not develop on a large scale.

Coal beds and intercalations of varying thickness are encountered in many localities of the Kuibyshev region. In the Samara Bend region, where the coal-bearing deposits contain large amounts of sooty siltstone, stigmarian mudstone and carbonaceous shale, coal beds and intercalations were discovered in the Syzran and Zhiguli areas, in the Strelny and Zolny ravines, and elsewhere.

This coal is usually brownish-black, dull, commonly banded (due to thin lenticular inclusions of vitrain), with irregular, acute-angled jointing, frequently argillaceous, developing into carbonaceous shale. The coal and carbonaceous shale, which are barely distinguishable from one another, sometimes attain a thickness of 3—4 m.

The Bobriki horizon in the Syzran region consists of well-rounded, largely quartzose sandstone alternating with clay. The sandstone and clay (especially the latter) contain very large amounts of carbonaceous material; if the latter predominates, the rocks develop into carbonaceous shale.

The coal-bearing deposits in the Yablonevy ravine, located 60 km to the east, differ little in composition from those in the Syzran section. A 10-m-thick member of carbonaceous clay with intercalations of siltstone and coal occurs at the base.

In the adjacent Morkvashi area the section is essentially different. Here it consists largely of clay-silt rocks with occasional thin intercalations of sandstone. As the section thins out to 20 m, its content of carbonaceous material also decreases perceptibly. Further east, in the Strelny ravine, the coal-bearing deposits also consist mainly of clay-silt layers with thin intercalations of carbonaceous shale and coal.

South of the Samara Bend, in the Pokrovsk area, the coal-bearing section is made up of alternating members of carbonaceous or sooty clay and siltstone and quartzose sandstone. The lithology of the members is variable, and the rocks grade into one another horizontally. The lack of persistence of the section, both vertically and horizontally, is a characteristic feature of the coal-bearing beds.

To the east and northeast of the Samara Bend is located a strip of thick alluvial deposits. In the area of Bolshaya Rakovka a coal bed about one meter thick was encountered in clays rich in carbonized plant remains. Fifty kilometers further north, in the Radayevka area, two coal beds were penetrated by well 1. The first bed, 1.2 m thick, was encountered at a depth of 1395 m, while the second, 2.5 m thick, is at a depth of 1402 m.

Between Bolshaya Rakovka and Radayevka are the Gorky ravine and Malinovka areas which correspond to the deepest part of an ancient river valley. Here the carbonaceous-argillaceous terrigenous member contains abundant carbonized plant remains, sometimes in the form of fragments of stems, leaves and bark. In the clay and siltstone there are, in addition to the accumulations of carbonaceous material, up to 8 coal intercalations, with a maximum thickness of up to 0.3—0.5 m. In southern areas of the Kuibyshev region, in the Mukhanovo area, well 7 passed through only one coal interlayer at a depth of 2070 m.

East and northeast of the ancient valley the thickness of the coal-bearing deposits decreases sharply. However, coal is also present here, e. g., in the Sernovodsk oil-bearing region where several wells penetrated coal seams, 0.3—1.6 m thick, at a depth of about 1300 m; coal was also found in the Uzyukovo, Borovsk, Yakushkino, Baitugan, Bezenchuk, Pokrovsk and other areas.

Coal occurrences are widespread in the northwestern part of Bashkiria. They are usually found in places where the coal-bearing horizon increases in thickness, owing to alluvial-fluviatile formations. The coal beds revealed by test holes extend for about 3—4 km, often pinch out and are interbeaded with arenaceous-argillaceous rocks. The number of coal seams in the section sometimes reaches 6 to 9; their thickness ranges from several decimeters to 24 m (well 361 in the Osha area).

One may judge the character of the coal deposits from the cross section through the Arlan area (see Figure 53). These deposits are typical lenticular formations of an alluvial valley, the formation of which was controlled almost exclusively by paleogeographic factors. The part of Bashkiria under discussion here was once the coastal zone of a Middle Visean sea. This probably explains why the areas richest in coal coincide in many cases with

oil-bearing areas, as has been noted at Bolshoi Arlan, Yangiznaratov, Ili-
shevo, Oryebashevo, Yugomashevo, Maksimovo, Chetyrmanovo and elsewhere.
The joint occurrence of coal and oil led Ovanesov and Yarulin (1960) to
suggest that these fossil fuels in northwestern Bashkiria are of common
origin.   They also believe that certain varieties of bituminous coals are
"some kind of pyrobitumen."

In the light of the available data, it is quite obvious that these views are
in error.   This follows, in particular, from the mineral composition of the
Lower Carboniferous coal of Bashkiria, which was studied by V. V. Kiryukov
at the laboratory of coal petrography of the Leningrad Institute of Mining.
He arrived at the conclusion that predominant among these coals are dull,
compact or friable, thinly bedded, homogeneous varieties which, according
to their microstructure, are spore durain and clarain-durain.   Somewhat
less common is dull bedded coal with lenticular inclusions of sooty macro-
components.   Judging from their microstructure, the latter are mixed xylain-
fusain-spore durain and clarain-durain.   Semidull striated and striated-
banded mixed durain-clarain and clarain coals are also fairly common.
Coal of a type intermediate between spore durain and cannel coal is of particular interest.

About 50 samples of coal from wells of the Arlan, Oryebashevo, Cheraulsk,
Urtaulsk and other areas were analyzed.   The characteristics of nearly all
samples, with the exception of their ash content, were found to be quite similar.   Therefore, we shall only present the extreme values recorded (in %):
2.7—5.8 moisture;  10.0—45.7 ash;  1.8—7.7 total sulfur;  37.4—53.9 volatiles
in the combustible mass;  coke button cakes slightly;  calorific value 3619—
—6962 cal.

The elemental composition determined for a small number of samples is
as follows (in %): 73.5—78.8 carbon;  5.7—6.1 hydrogen;  nitrogen + oxygen
16.4—17.3%;  density 1.14—1.77 g/cm$^3$.

Bashkirian coal differs little from that of the Ural-Volga region and also
belongs to the inferior bituminous coals, i. e., long-flame coal sometimes
approaching gas coal.   It should be stressed that the majority of the coal
samples studied possess some features of brown coal and are often classified as transitional from brown to bituminous.   Both the coal and the Lower
Carboniferous oil in Bashkiria have a high sulfur content associated with
the geochemical environment in which the producing beds were formed.

The considerable content of sulfur and its compounds, both in the coal
and the enclosing rocks, points to the stagnant nature of the water bodies
with prevalence of reducing or strongly reducing environments in the benthic
zone.   As these water bodies became overgrown with peat bogs, a certain
portion of the organic remains in the upper layers of the latter became
oxidized, while the water near the bottom became enriched in carbon dioxide
due to the intensive consumption of dissolved oxygen.   The growth of the
peat bogs was accompanied by the development of sulfate-reducing bacteria.
The latter were apparently responsible for the formation of pyrite from the
sulfates of swamp waters saturated with ferruginous solutions.

We shall briefly discusss the genesis of sulfur in coal in general and of
organic sulfur in particular.   Sulfur is known to be present in coal either in
the form of sulfide or pyrite and sulfate, or as organic compounds, the first
type clearly dominating in high-sulfur coal.   Sulfate sulfur occurs mainly

as calcium sulfate $CaSO_4 \cdot 2H_2O$ and less commonly as ferric sulfate $Fe_2(SO_4)_3$, or ferrous sulfate $FeSO_4$. The second type includes organically bound sulfur in the form of complex organic compounds. Its content is calculated as the difference between the total sulfur and the sum of sulfate and sulfide sulfur. As reported by Gebler (1959), the content of organic sulfur in the coal varies widely: from 25 to 50% in high-sulfur coal to as high as 80% in low-sulfur coal.

Gebler explains the origin and character of the organic sulfur in coal as follows. First of all, let us consider that part of the sulfur whose origin is associated with the proteins of coal-forming plants. Its amount is very small. If one assumes that wood was the main material from which the coal was produced, then one must bear in mind that present-day wood only contains 0.005% sulfur. Assuming that all this sulfur passed into the coal while the proportion of carbon increased from 44.4 to 90%, its content in the coal should be about 0.009–0.01%. However, the actual content of organic sulfur in coal is many times greater. Moreover, part of the sulfur was apparently lost to the atmosphere together with water during hydrogen-sulfide fermentation. Consequently, the formation of sulfurous organic substances in the coal bed is due to the elemental sulfur which is liberated during the reduction of ferrous sulfate by hydrogen sulfide, according to the following reaction:

$$2FeSO_4 + 5H_2S = 2FeS_2 + 2S + H_2SO_4 + 4H_2O.$$

Therefore, the above reaction, which is primarily due to the actual geochemical conditions, leads to the formation of pyrite, elemental sulfur and sulfuric acid. Elemental sulfur is used by bacteria for the formation of hydrogen sulfide, while the reaction between sulfuric acid and the calcium carbonate of the enclosing rocks and within the coal bed proper may produce gypsum.

Strakhov et al. (1959) attribute the rather high sulfur content and the wide occurrence of pyrite in the Middle Visean coal-bearing deposits to the considerable amounts of organic matter buried within the sediments. This led to the rapid development of pyrite from the authigenic mineral forms of iron and the suppression of carbonate-leptochlorite formation.

The pronounced pyritization of the coal-bearing rocks and the high content of sulfur in the coal enable us to appraise the inland water bodies of the Middle Visean. In these water bodies, as well as in the marine basins, microbiological reduction of sulfates took place. These processes resulted in the binding of sulfur with iron to form pyrite. The inland water bodies of that time apparently differed appreciably from present-day ones in that they were fed by rivers whose waters were richer in suspensions due to intensive chemical weathering and, especially, to the abundance of humic substances in the rivers. Strakhov believes that these substances serve partly as stabilizers of colloid solutions of mineral salts and partly as their direct carriers.

Reverting now to the description of coal occurrences as a criterion for determining the paleogeographic environment, it is necessary to point out the general pattern noted for the deposits under study. This pattern has resulted from the fact that rather thick, single-layer coal deposits usually

accumulated in peneplaned areas of the ancient continent, while in river
valleys and deltaic regions the number of coal beds increases, but they are
thinner, have a high ash content and are not persistent. This is borne out
by the comparison shown in Figure 58, as well as by the petrographic composition of the coal.

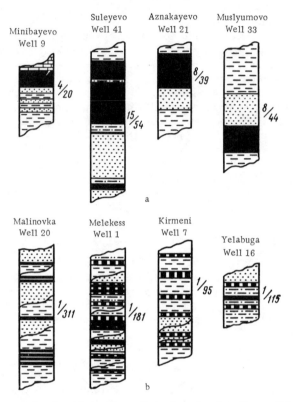

FIGURE 58.  Characteristics of coal beds of the producing formation of the Yasnaya Polyana substage
in the Volga-Ural region, formed on a plain (a) and in a river valley (b).  The numerators of the fractions to the right of the columns indicate the thickness of the coal bed in meters, while the denominator
gives the thickness of the coal-bearing formation (in meters).

*Lower Reaches of the Volga*

In this part of the Volga region the coal-bearing Middle Visean deposits
were penetrated by wells mainly in the Saratov and Volgograd areas where
they are virtually restricted to the right bank, on a vast accretion plain.
Coal was noted in core samples from oil test holes drilled in the Teplovka, Yelshanka, Irinovka, Balanda, Sokolovogorsk, Surovo, Peschany Umet,
Bogayevo, Goryuchnino, Ilovlya, Bakhmetyevka, Linevo, Zhirnovka, Korobki
and Archedino regions, and also elsewhere in the lower reaches of the Volga.
However, it should be borne in mind that correct determination of the rich-
ness and the true thickness of the coal seams depends both on the density
of the boreholes and the specific drilling techniques employed.  The fact

that the average recovery of core samples did not exceed 10% adds to the problem and helps explain why the thickness of the coal deposits has been underestimated.

The coal samples selected from cores from the above-mentioned bore-holes are in most cases macroscopically compact, sometimes thinly bedded, dull, semidull, less frequently semibright, with an irregular jagged fracture, brown or brownish-black. They are undoubtedly autochthonous. Only in cases when the development of the peat bogs was interrupted by meandering stream channels or by sudden flooding, could a certain rewashing and redeposition of plant remains with large amounts of mineral material take place, leading to the formation of high-ash coal or more commonly carbonaceous rocks.

Petrographic and chemical studies of the coal encountered in different areas and at different depths in the Lower Volga region showed that it is generally similar to that of the Middle Volga and Kama regions. According to microscopic determinations, most of the coal of the Volga region near Saratov and Volgograd is humic, and only a small amount is humic-sapropelic and sapropelic-humic. According to petrographic studies (phyterals and macerals), the coal is of the durain and clarain type. The former accumulated in exorheic lakes and swamps where it became enriched in spore exines, cuticles and other more stable plant fragments. The clarain varieties, consisting of macerals and approximately equal amounts of terrestrial plant remains and microspores, were largely formed in quagmires where the process of jellification was especially complete.

Individual coal samples from the Yelshanka, Balanda and Bakhmetyevo oil- and gas-bearing areas contain fusain. This suggests that in this region the swamps either dried up or became shallower and the plant remains were oxidized.

Some boreholes passed through intercalations of coal containing algae. Petrographically they are mostly clayey microspore durains containing *Pila, Cladiscothallus* and *Reinschia.* The presence of the clayey material in this coal is indicative of a supply of water to their sites of accumulation. These were, in all probability, water bodies of the relict lake type or limans which were separated from marine lagoons, in which algae could develop alongside higher plants.

The carbonaceous rock group consists of high-ash coal rich in mineral impurities. This coal has lost its principal features and is transitional to carbonaceous mudstone or carbonaceous siltstone.

Consequently, the petrography of the coal is directly dependent on the conditions under which the primary source material accumulated. Further qualitative change of the primary source material from which coal was produced, toward carbonization, occurred under the influence of biochemical processes, mainly during the period of diagenesis prior to the formation of the roof and the subsequent metamorphism. The extent of coalification, and therefore the physical, chemical and technologic properties of the coal are controlled not only by the general character of its accumulation, but also by the subsequent metamorphic processes.

At the time when this territory was under platform conditions, only regional metamorphism could have taken place. Its decisive factors are the pressure of the overburden and the temperatures at the depth of the coal-bearing beds.

Characteristics of Coal

The analytical results presented in Table 7 provide an indication of the extent of coalification of the Middle and Lower Volga coal. A notable feature is the carbon content and the amount of volatile matter given off on combustion of a unit mass, as well as the relatively high calorific value. According to most of these characteristics the coal is closer to bituminous than to brown coal varieties, although it is difficult to draw a sharp dividing line between them. Incidentally, the terms "brown" and "bituminous" are rather vaguely defined, and an entire gamut of transitional varieties exists between them. The transition of fossil coal from the "brown coal" stage to the "bituminous" stage is evident not only from the elemental composition, calorific value and other features, but also from the fact that in most cases the coal lacks soluble humic acid, and is only slightly hygroscopic, i.e., it has a low moisture content (on the average about 4%).

TABLE 7.  Results of elemental analysis of coal from the Middle and Lower Volga regions

| Area | Well No. | Depth of sampling, m | Moisture (analytical), $W^a$, % | Content in dry material, % | | Content in combustible mass, % | | | Calorific value $Q^c$ |
|------|----------|---------------------|------|------|------|------|------|------|------|
| | | | | ash $A^d$ | sulfur $S^d$ | volatiles $V^c$ | carbon C | hydrogen H | |
| Golodyaevka .......... | 9 | 1112–1121 | 2.68 | 32.35 | 4.70 | 51.53 | 76.71 | 4.42 | 7698 |
| Gorky ravine ........ | 23 | 1528–1533 | 1.38 | 18.67 | 2.00 | 60.68 | 76.03 | 6.60 | 8467 |
| Radayevka ............ | 1 | 1420–1426 | 3.21 | 13.59 | 3.45 | 55.64 | 76.85 | 5.48 | 8067 |
| Krasnovka ............ | 1 | 957–963 | 7.43 | 29.65 | 6.33 | 47.85 | 71.52 | 2.51 | 6809 |
| Yelabuga ............. | 18 | 1021–1031 | 9.69 | 9.97 | 2.75 | 43.38 | 75.18 | 3.20 | 7406 |
| Suleyevo............. | 41 | 1053–1059 | 2.90 | 14.51 | 4.94 | 43.73 | 75.66 | 4.65 | 7748 |
| Aznakayevo .......... | 21 | 1063–1066 | 6.00 | 15.42 | 2.81 | 40.98 | 75.20 | 3.55 | 7542 |
| Mouth of the Kama .... | 3 | 939–948 | 5.53 | 13.70 | 3.89 | 48.90 | 72.05 | 3.30 | 7000 |
| Kirmeny ............. | 9 | 1000–1006 | 6.75 | 26.95 | 1.98 | 38.29 | 73.61 | 2.74 | 6878 |
| Saraily ............. | 3 | 1170–1180 | 3.92 | 16.46 | 2.10 | 43.60 | 75.59 | 4.75 | 7582 |
| Balanda ............. | 2 | 1011–1018 | 8.33 | 29.10 | 2.76 | 46.83 | 69.52 | 6.39 | – |
| Yelshanka .......... | 223 | 841–848 | 4.01 | 18.48 | 2.87 | 46.58 | 73.92 | 5.50 | 7472 |
| Peschany Umet ....... | 9 | 1074–1082 | 5.80 | 21.22 | 5.63 | 49.67 | 71.25 | 5.93 | 7454 |
| Zhirnovka ........... | 15 | 1036–1043 | 3.42 | 17.75 | 5.04 | 51.30 | 75.85 | 6.02 | 8000 |
| Linevo ............ | 4 | 1275–1280 | 2.23 | 14.38 | 4.63 | 56.10 | 77.70 | 6.95 | 8412 |
| Korobki ............ | 34 | 1270–1274 | 1.07 | 19.03 | 3.74 | 47.69 | 68.71 | 6.14 | 7066 |
| Ust-Buzuluk .......... | 39 | 254–258 | 9.65 | 18.15 | 2.07 | 46.69 | 73.20 | 5.15 | 7315 |

The yield of volatiles is rather high, sometimes exceeding 60%. However, the change in the content of volatiles does not always obey Hilt's law, according to which the coal rank increases with depth. To cite an example, in the thick coal bed encountered in Suleyevo well 41, the volatile content decreases noticeably with increasing depth, and over the depth interval of 1071—1081 m it amounts to 33%. However, the yield of volatiles from coal samples from other boreholes, e. g., the Gorky ravine 23 well for the depth interval of 1528—1533 m, exceeds 60%, and the coal from the test hole in the Melekess area, depth interval 1597—1601 m, gives off up to 50% volatile matter on combustion.

The carbon content of the coal being studied ranges from 64 to 82% and averages about 70%; the degree of coalification generally increases with depth, although there are considerable deviations in both directions.

All these deviations from Hilt's law must be due to differences in petrographic composition, the content of mineral impurities (ash) in the coal, and the pressures and temperatures at the time of coalification. Some idea of the "maturity" or degree of coalification can be obtained from the density of the coal, which seldom exceeds 1.15—1.20 g/cm$^3$, corresponding to brown coal or relatively low-rank bituminous coal.

One of the most significant qualitative characteristics of coal is its ash content. The tables show that it varies within fairly broad limits, from 8% in relatively rare, pure varieties to more than 40—45% in rather widespread carbonaceous shale (carbonaceous mudstone or siltstone). Thin lenticular deposits are the lowest in ash. This range of ash contents indicates the diversity and frequent change of conditions of coal accumulation, primarily due to paleoclimatic features.

According to its content of total sulfur (on the average, about 3—4%), the coal is generally one of the high-sulfur varieties and is similar in this respect to that of the Kizel basin. Analytical results show the presence of phosphorus in trace amounts, not exceeding 0.01—0.02%.

Thus, the combined physicochemical properties of this coal indicate that it is transitional from brown to bituminous, fairly close to long-flame, and in certain sections (particularly deep ones) it is similar to gas coal. The degree of coalification generally increases toward the east.

If one judges the degree of metamorphism of the enclosing rocks from the rank of the coal, one can conclude that the enclosing rocks, which were subjected to general regional metamorphism only, retained properties quite favorable for the accumulation and preservation of large oil and gas pools.

### Coal as an Indicator of the Degree of Catagenesis

Coal is a rather objective indicator not only of humid-continental conditions of sedimentation, but also of the rank or degree of metamorphism (catagenesis) of the coal body proper and of the enclosing rocks. Since the coal-bearing beds of the Yasnaya Polyana substage in many parts of the Ural-Volga region also contain oil and gas, it might be of interest to specify the effect of regional metamorphism on the distribution of oil and gas pools.

This is also merited in view of the fact that petroleum geologists consider the degree of lithification or alteration of sedimentary rocks as one of the criteria for assessing oil and gas prospects. It is common knowledge that highly metamorphosed rocks virtually do not contain oil deposits and that economic accumulations are not encountered in primary Recent sediments, even in cases when the latter are in the stage of diagenesis.

One of the methods of determining the degree of lithification of rocks is to establish the carbon ratio of the coal, which is expressed as the percentage of fixed carbon in the coal. The derived values of this ratio usually range from 20% for coal of the lowest rank to 95% for anthracite. In other words, the higher the carbon ratio the stronger the metamorphism. E. P. Lilley has reported that most of the oil fields on the North American Platform occur in deposits with an average carbon ratio ranging from 55 to 65%.

Another indication of rank is the reflectance of vitrinite, which is determined under the microscope on polished sections. Both methods of determining the rank of coal call for special studies. However, a qualitative characterization of the degree of alteration of the coal due to metamorphism is quite sufficient for drawing preliminary conclusions as to the possibility of formation of oil or gas pools in certain coal-bearing complexes. This qualitative characterization is a very good index of the rank of coal in which the yield of volatiles regularly decreases and the content of carbon increases from brown coal to anthracite varieties (Table 8).

TABLE 8. Dependence of yield of volatile matter on coal rank

| Rank of coal | Elemental composition, % | | | Yield of volatile matter, % |
|---|---|---|---|---|
| | C | H | O | |
| Peat ..................... | 58.0 | 6.0 | 33.0 | 70 |
| Brown coal (B) ........... | 70.0 | 5.0 | 25.0 | 53 |
| Long-flame coal (L) ....... | 71.0 | 5.5 | 12.5 | 43 |
| Gas coal (G) .............. | 81.0 | 5.4 | 8.3 | 39 |
| Steam-fat coal (SF) ........ | 83.0 | 5.1 | 5.1 | 32 |
| Coking coal (C) ........... | 87.0 | 4.8 | 3.6 | 23 |
| Steam-caking coal (SC) ..... | 89.0 | 4.5 | 2.7 | 16 |
| Lean coal (L) ............. | 90.0 | 4.2 | 2.3 | 12 |
| Anthracite (A) ............ | 95.0 | 1.8 | 1.8 | 3.5 |

It is quite evident that one should not attach universal significance to either the carbon ratio or the reflectance of vitrinite, especially in areas where oil and gas are known to occur, because the very presence of these pools is in itself indicative of the favorable degree of metamorphism of the producing formation.

The effect of regional metamorphism on the occurrence of oil and gas and the implications of the carbon ratio theory are discussed in detail by Russell (1958).

The most significant assumption of this theory is that oil is rarely present in localities where the carbon ratio rises above a certain value. However, as acknowledged by Russell himself, there are still too many gaps in our knowledge of the significance and reliability of the carbon ratio theory. The same applies to vitrain alteration (its reflectance) as an indication of the degree of metamorphism, which may be used for assessing the reserves of oil and gas.

R. S. Tarr refutes the significance of the carbon ratio and, consequently, of metamorphism in oil exploration. He maintains that the absence of oil or the replacement of oil by gas in areas where the carbon ratio exceeds 60% is the result of variations in the primary conditions of sedimentation and is not due to changes in the degree of regional metamorphism. Russell offers rather convincing arguments in support of Tarr's viewpoint. In southeastern Oklahoma, for example, there are regions covered by thick sediments which contain no petroleum whatsoever, notwithstanding the fact that the carbon ratio there is much lower than 60%.

If one now considers the deposits of Yasnaya Polyana age, one readily discerns a similar pattern on comparing the sections of the Lower Volga region and the Pripyat graben. The terrigenous beds within the latter are in places more than twice as thick as those in the Volgograd region. According to coal analyses, the degree of regional metamorphism in the Pripyat graben would appear to be even more favorable for the formation of oil deposits than that in the Lower Volga region. However, the physiographic conditions of sedimentation in the Pripyat part of the East European Platform were such that no oil could be formed there.

Consequently, if a qualitative analysis of the coal is used as a criterion for determining the degree of regional metamorphism and the latter's influence on the occurrence of oil and gas, it is essential to take into account the actual paleogeographic environment in which the source beds and reservoir rocks were formed.

We shall not discuss here the organic matter of the coal as an indicator of the degree of lithification of sedimentary rocks during catagenesis. This aspect has been covered sufficiently in several specialized works and is hardly relevant to paleogeography.

## 12.  *THE HYDROGEOLOGIC FACTOR IN OIL AND GAS ACCUMULATION AND PALEOGEOGRAPHY*

We have discussed the paleogeographic environments of deposition of the Lower Carboniferous terrigenous beds and the influence of these environments on the composition and distribution of sediments containing fossil fuels, as well as the dependence of the formation of oil pools on the nature of and relationships between the source beds and reservoir rocks. The formation of oil deposits would be impossible without the active participation of water. The migration and accumulation of mobile hydrocarbons are controlled, among other factors, by subsurface water.

The most productive oil- and gas-bearing zone of Middle Visean sediments extends not only along an ancient coastline, but also along the present-day margins of the East European Platform. The latter pass into foredeeps where there is a system of artesian basins with water under hydrostatic pressure. This is of utmost importance for the migration of hydrocarbons from the source beds to zones favorable for their accumulation.

The role of hydrogeologic factors in the formation, re-formation, preservation and destruction of oil fields has been treated by A. I. Silin-Bekchurin,

N. K. Ignatovich, M. A. Gayal'skii, V. A. Krotova, I. O. Brod and many others. In particular, Silin-Bekchurin (1948), while analyzing the paleohydrogeologic conditions in the Ural-Volga region, attributed the movement of subsurface water and oil to changes in the structural-tectonic pattern. He believes that the tectonic processes disrupted the hydrodynamic and geochemical equilibrium of the underground water, displacing regions of recharge and drainage and creating conditions in which waters flowed from structure to structure or to newly formed drainage zones. He points out that both uplift and subsidence of the territory caused migration of oil: subsidence results in migration of the oil along the strata, whereas uplift leads to its migration along fissures between them.

According to Silin-Bekchurin, migration in the Ural-Volga Paleozoic rocks took place along each oil-bearing stage separately, except for those deep-seated dislocations along which underground water could move from one oil-bearing stage to another within different localized sections. As follows from the available data on oil finds, this was also the case with regard to the Lower Carboniferous terrigenous beds under study.

The foredeep region, which was filled with the sediments of a shallow sea during the Lower Carboniferous, later developed into a region of separate, closed or interconnected artesian basins (Caspian, Ural, Dnieper-Donets, etc.). These were extensive downwarped crustal areas which were either confined between the gently sloping platform margins and mountain structures or were located between large uplifts of the platform. The mountain structures and positive structures of the first order, which were uplifted considerably relative to the platform, were zones of recharge which created the necessary hydrostatic head to cause the migration of oil from the source beds in depressions into the reservoir rocks surrounding these depressions. It is now quite evident that a special role was played by alluvial-deltaic and fluviatile formations of ancient river valleys which, cutting through the rims of the basins, drained them and at the same time served not only as convenient channels for the migration of fluids, but also as good reservoirs for the accumulation of liquid and gaseous hydrocarbons.

The estuarine part of the valley of the paleo-Kizel is a good example of such a situation. The oil content of the Carboniferous terrigenous rocks there increases in the proximity of the floodplain portion of the ancient valley, on the one hand, and toward the deltaic zone, on the other. For instance, while the Yasnaya Polyana substage in the Krasnokamsk, Severokamsk and Nytva areas contains only one petroliferous sandstone bed ($B_2$), there are two such beds ($B_1$ and $B_2$) in the Polazna area, and three ($B_1$, $B_2$, and $B_3$) in the Yar deposit. The oil-rich Yar sandstones (thickness exceeding 64 m) in the Kamennolozhskaya area, which is a continuation of the Yar area, are even closer to the platform margin and to the central part of the fluviatile deposits along which subsurface drainage took place.

The distribution of the Lower Carboniferous oil deposits along the ancient shore of the Middle Visean sea meets not only certain paleogeographic and tectonic requirements, but also hydrogeologic requirements, since the land-sea boundary passed through the zone of transition from the platform to the foredeep, which consequently became an artesian basin. Krotova (1962) concluded that in zones where the platform adjoins downwarps beyond which there are regions of folding, hydrodynamically stressed systems are created in which the descending movement of the waters in the near-platform margin of the depression changes to an ascending one, owing to the differences in the heads of the waters in the recharge zone and at the margin of the platform. It is in the areas adjacent to the depression, with an ascending movement of the waters, that oil migrates from the depression and forms pools, as suggested by Krotova. We need not go into a detailed discussion of the role of subsurface water in the formation of oil and gas pools, since this subject has been discussed in numerous specialized works. We shall only mention that most hydrogeologists, in relating the movements of underground water to the tectonic structure of the earth's crust, tend to underestimate the importance of the lithology and, especially, the actual structure of the sedimentary strata along which this movement takes place.

Although we ascribe the major role in the change and shifting of regions of recharge, runoff and discharge of subsurface water, and, therefore, the character of movement of the fluid masses, to the geotectonic factor, the significance of the paleogeographic factor should not be overlooked. The most convenient path or channel for the lateral migration of fluids from the basins to the marginal zones of the platform was via porous and permeable sediments whose formation was controlled mainly by the paleogeographic sedimentary environment.

If the fluid moves in the direction of flow of the subsurface water, i. e., from the regions of recharge to those of runoff and discharge, the zone of oil and gas pools should generally coincide with the latter. Such is the case along the margins of the East European Platform where broad erosion valleys of large paleorivers, filled with alluvial deposits, were favorable sites for runoff and discharge, as well as for the separation of the fluid into its components: water, oil and gas.

This separation of oil and gas has nothing to do with the layered free flow of gaseous and liquid hydrocarbons over long distances, as assumed by certain investigators (A. L. Kozlov, V. F. Lipetskii and others). Brod (1960) rightly noted that "their assumption about the migration of hydrocarbons in the form of a continuous phase (oil or gas flow) from the Caucasus and Urals to the East European Platform harks back to the 19th-century fantastic concepts of subterranean petroleum rivers linking oil- and gas-bearing regions of different countries." Brod stressed that it is always necessary to consider hydrocarbons and their phase state in conjunction with water, bearing in mind that their amount is negligibly small relative to the mass of water. One should also recall that hydrocarbons move in the aqueous solution and are separated from it following a change in the physicochemical environment.

Some investigators (O. A. Radchenko, M. A. Kapelyushnikov, T. P. Zhuze and others) are of the opinion that the migration of hydrocarbons over considerable distances is possible only in the gaseous phase. According to Krotova (1962), methane is the main gaseous component of water in oil- and gas-bearing regions. It is usually accompanied by heavy hydrocarbons. The rather considerable increase in the pressure of water-soluble gases nearer the foredeeps is an additional proof of hydrocarbon migration from depressions to the platform, i. e., from high-pressure to lower-pressure regions. During the course of their migration the waters saturated with hydrocarbon gases, on cooling down and entering a zone of lower hydrostatic pressures, release gases and fill the traps.

The fact that petroleum gradually becomes lighter nearer the foredeeps and begins to change into condensate, is interpreted by Krotova as a gradual change in the phase state of the hydrocarbons; this is indicative of its movement in the direction of increasing density of the fluid. The fluid, on entering an environment of lower temperatures and pressures and losing its light fractions on the way, gradually acquires the properties of a liquid phase and forms oil pools. The above assumptions are in good accord with the so-called differential migration theory (or the theory of consecutive accumulation of gas and oil). We fully agree with Krotova that hydrocarbons can migrate in the liquid phase over relatively small distances. However, we cannot share her opinion that the migration of hydrocarbons from deep depressions fringing platforms does not in any way contradict the inorganic origin of oil, because deep faults exist not only on the platform but also in the depressions. The results of lithofacies, geochemical and paleogeographic analyses of the Lower Carboniferous producing formation disprove the hypothesis of the inorganic origin of the oil; these data also help to explain why regional depressions or basins form extensive zones of oil accumulation.

Whereas arenaceous rocks occurring among argillaceous or carbonate beds serve as the main channels for the lateral migration of fluids, the thick alluvial-deltaic paleoriver formations in the lithostratigraphic complex being studied only widen the path and increase the extent of this migration. All the detected river paleovalleys are oriented almost at a right angle to the cis-Ural depression and the Caspian basin, whence the hydrocarbons migrated into pools located along the margins of the East European Platform.

The spatial disposition of oil pools of the Lower Carboniferous terrigenous beds points to their association with the coastal zone of the Middle Visean marine basins, which, in their turn, were confined to regional, closed or semiclosed sections of the downwarped crust. The oil pools themselves are controlled by local structural or lithologic traps which are parts of the general system of a particular artesian basin. The oil and gas filling the traps are in equilibrium with the water of the enclosing rocks. This equilibrium is controlled by the relationship between the pools, on the one hand, and the head and direction of the groundwater runoff, on the other.

Thus, the general hydrogeologic environment of the Lower Carboniferous producing formation of the oil- and gas-bearing territory of the East European Platform is undoubtedly related not only to the structural pattern, but also to the presence of buried valleys of paleorivers and their deltas. The hydrodynamic regime, as well as the migration of hydrocarbons, was controlled to a considerable extent by the distribution pattern, thickness, persistence, permeability and other properties of the numerous arenaceous formations within the Visean terrigenous beds.

Therefore, a detailed reconstruction of the paleogeographic environment in which the productive complexes were formed may help to determine the paths and directions of fluid movement, such knowledge being essential for an understanding of the formation of oil and gas pools.

## 13. PALEOGEOGRAPHIC ZONALITY IN THE DISTRIBUTION OF OIL AND GAS POOLS

In the preceding chapters of this book we have stressed the fact that numerous oil and gas fields occur in the coastal zones of ancient seas, the largest deposits being associated with regions of deltaic and prodeltaic sediments. Is this merely a coincidence or does the distribution obey a certain law?

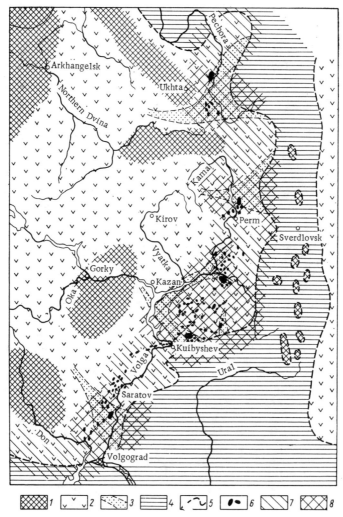

FIGURE 59. Location of major oil deposits in the Yasnaya Polyana producing formation of the Ural-Volga region; belts and major zones of oil and gas accumulation:

1 — elevations of the ancient relief; 2 — lowland plain; 3 — river valleys and their mouths; 4 — marine basin; 5 — shoreline; 6 — oil deposits; 7 — oil and gas accumulation belt; 8 — major zones of oil and gas accumulation.

We have already attempted to answer this question when considering the relations between dry land and sea, between physiographic environments and the evolution of the biosphere, and between the formation of source beds and reservoir rocks.  Awareness of the above relationships enables one to discern a most specific paleogeographic zoning of oil and gas accumulation, which follows from the general distribution pattern of fossil fuels.

We shall try to confirm this conclusion, once again using as an example the distribution of oil deposits in the terrigenous beds of the Yasnaya Poly- ana substage, the physiographic depositional environment of which has been studied in fairly great detail.  Let us plot the major oil deposits discovered in these beds on a paleogeographic map (Figure 59).  One notes that nearly all of the deposits are grouped along the coastal zone of the marine basin which existed at the time when these sediments accumulated.  Let us con- sider the distribution features of the last-mentioned oil pools, primarily, those with large reserves.

The richest Lower Carboniferous oil pools discovered in the Ural-Volga region to date are those in the southeast of the Kuibyshev region and in the north of the Orenburg region.  Near Kuibyshev, in the zone of the vast delta of the Middle Visean paleo-Kama, large commercial oil deposits occur in the sandstones of the terrigenous beds.

Toward the north, along the Kama-Kinel depression, i. e., up the valley of the paleo-Kama, in the region of the Gorky ravine, Malinovka and Radayevka, the yield of the oil wells decreases.  No oil was discovered in the Melekess test hole further to the north.  This implies that the source-bed region was on the continental shelf of the marine basin, into which the paleo-Kama de- bouched.  This is why the prodelta sandstone is more saturated with oil than the fluviatile sandstone at a greater distance from the shelf.  Yet another conclusion suggests itself: during Middle Visean times the Kama-Kinel de- pression was not the region where the source beds were formed.

The Carboniferous oil from the Mukhanovo deposits has a high density, a high content of tar, sulfur and paraffin, and a relatively low yield of light fractions (benzene and kerosene).

The high productivity of the Mukhanovo and neighboring oil deposits is due, firstly, to their occurrence in the deltaic sediments of a huge paleo- river and, secondly, to their location on the rather steep southeastern slope of the East European Platform.  It was here, in the nearshore and particu- larly the estuarine parts of the sea, that the enormous masses of river water draining from the continent changed the environment radically; mechanical, chemical and biological processes of sedimentation were especially vigorous. Here, in all probability, the primary-bituminous source beds were formed, from which oil migrated, in the diffuse state, to higher horizons in the ad- jacent areas where it accumulated in suitable structural or lithologic traps.

In the adjacent Orenburg region about 62% of the oil is produced from Lower Carboniferous sediments.  The main producing horizons are the coal- bearing terrigenous and, partly, the Upper Tournaisian carbonate complexes. The oil deposits in the Krasnoyarsk, Zaglyadino, Sultangulovo, Tarkhanovo and other regions in the central part of the Bolshoi Kinel arch also occur in deltaic sediments.  It is not by chance that the Bobriki sanstones here have the form of narrow long strips extending in a submeridional direction.  It

has been established by drilling that these strips continue fairly far to the
north.  As reported by Kulakov, Svishchev and Panteleev (1962), the channel
sands in the markedly downwarped Bolshoi Kinel arch broaden and form
detrital fans of terrigenous material or peculiar "microdeltas."  The sand
deposits have a porosity of 17—25% and a permeability of 0.3—2.0 darcies.
The oil wells generally have high yields.

Seismic surveys and drilling data indicate that the oil pools in the Middle
trans-Volga region are also associated with deltaic or fluviatile deposits.
The diagram of Elanskii and Tolkachev (1959) showing the distribution of the
maximum thickness of the Lower Carboniferous terrigenous beds suggests
that they bifurcate along two peculiar river courses, reflecting, in their
opinion, the pattern of ancient intraformational erosion.  The deltaic channels and tributaries first cut and deepened their courses, and then, as the
base level changed, the latter were filled with predominantly sandy material.

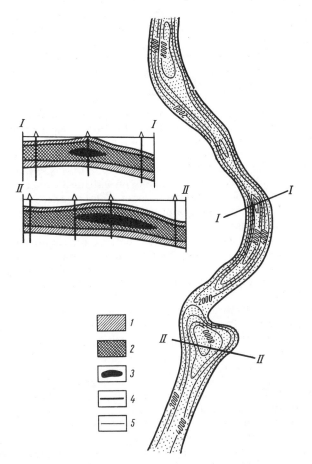

FIGURE 60.  Producing sand layer of the Pokrovskoye oil field (Morgunov et al., 1960):

1 — limestone; 2 — argillaceous rocks; 3 — petroliferous sandstone; 4 — boundaries of
the layer; 5 — piezoconductivity contours.

In the Pokrovskoye oil field located in the Chapayev district of the Kuibyshev region, a shoestring oil pool (Figure 60) was discovered in the Tula
horizon by drilling. It is a strip of fine-grained quartzose sandstone within
clays and extends in an approximately meridional direction. The length of
the pool is 10 km and its width ranges from 250 to 1300 m. The thickness of
the sandstone along its longitudinal axis increases from north (5 m) to south
(10 m). The sandstone pinches out toward the edges of the strip, where it is
replaced by clay. It is well sorted, has an average porosity of 20.6% and a
permeability of 0.75 (according to the core samples) or 0.8 darcies (according to production data).

The above features, together with piezometric results, suggest that this
shoestring pool is of the fluviatile type. The piezometric contours show
that the pressure had increased in the central part of the river course,
where the current was stronger and coarser sand was deposited. The equi-
potential lines suggest that the flow was from north to south.

In Bashkiria the main areas where the terrigenous beds of the Yasnaya
Polyana substage contain oil are situated in the Birsk saddle (Arlan-Dyur-
tyuli oil-bearing zone) and in the Tuimazy and Shkapovo regions. The productivity of the terrigenous beds is closely related to their thickness. An
increase in the latter, generally due to the presence of sandstone, is usually
accompanied by an increase in oil content.

We have already noted the obviously fluviatile origin of the sandstones in
the coal-bearing beds of the northwestern and southwestern parts of Bashkiria where oil and coal are encountered in the same deposits of the coal-
bearing horizon. Oil generally occurs higher up in the section than the coal.
It is of the heavy and high-sulfur variety, having a density of 0.88—0.91 g/cm$^3$,
rising to 0.92—0.93 g/cm$^3$ in small pools at Shkapovo. Its tar content is
12—18%, and sulfur content 2.5—3.5%; hydrogen sulfide is commonly present.

The distribution of the petroliferous sandstones is controlled by the structure of the sand bodies which are typical fluviatile strips consisting of sandstone lenses of varying thickness. This is also evident from the cross section of the Arlan field (see Figure 38) and from the diagrammatic map showing the distribution of sandstones in the southwest of Bashkiria (see Figure 39),
where they form fairly long but narrow strips. The latter stretch for tens
of kilometers in the direction of the general slope of the southeastern margin of the platform toward the Caspian syneclise.

It is noteworthy that Khalimov (1960) reported the discovery of both
structural and lithologic traps within the sand strips of the coal-bearing
beds in the southwest of Bashkiria; the lithologic traps are associated with
updip pinch-outs of the sandstone. Largest in size and reserves are two
lithologically sealed pools penetrated by several wells in the Shkapovo area.
In describing the Shkapovo paleoriver of Middle Visean age, we have already
mentioned the presence of three sand strips (river channels) apparently coinciding with the distributaries in the estuarine part of this river. Khalimov
believes that within the central, so-called Belebei-Shkapovo strip the hydrocarbons, which had migrated updip and impregnated the traps encountered
on their way, must have accumulated even in the most elevated northern part
of the strip sealed on all sides by impermeable rocks. This explains why
the sandstones of the coal-bearing beds in some parts of the Usen-Ivanovskoye

area contain lithologically sealed pools.  There are prospects for the discovery of new lithologic oil and gas pools associated with alluvial-fluviatile and deltaic sediments not only in this area, but also in the southwest of Bashkiria as well as in the adjacent territories of the Orenburg region.

If one assumes that the oil discovered in the Shkapovo river paleovalley originated from nearshore-marine source beds of the basin on the southeastern margin of the East European Platform, how does one trace the source from which oil could have migrated into the deposits in the Birsk saddle? This question arises since we regard the alluvial-fluviatile deposits of northwestern Bashkiria as belonging to one of the tributaries of the paleo-Kama, whose valley extended in a direction opposite to that of the ancient shoreline. In this particular case, however, one should take into account that the upper reaches of this tributary were in close proximity to the seashore.  Incidentally, in this respect the paleo-Kama resembles the present-day Niger which has its source near the Atlantic shore, penetrates deep into the continent, describing a large arc, and then debouches into the Gulf of Guinea.

On analyzing the distribution of the rich oil fields in the sand deposits of the paleo-Kama, one notes that the largest concentration of oil is encountered in its estuarine portion and upper reaches, i. e., parts which were either directly adjacent to the ancient marine basin or close to it.  Notwithstanding the considerable rearrangement of the structural pattern subsequent to the deposition of the Lower Carboniferous terrigenous beds, the original juxtaposition of the source beds and reservoir rocks must have played the decisive role in the formation of the oil deposits.

The richest oil pools in the Lower Volga region occur in areas where the thickness of the Lower Carboniferous terrigenous beds increases.  As already mentioned, one such area is the fairly extensive territory of the Ryazan-Saratov paleodelta whose cross section is shown in Figure 61.  Here several oil and gas fields were discovered within the Yasnaya Polyana substage (Korobki, Zhirnovo, Linevo, Bakhmetiyevka, and other fields).  Particularly rich in oil is the Bobriki horizon, although the oil occurs in carbonate rocks of the Tournaisian stage, as well as in sandstones of Tula age.

On the basis of the changes in thickness of the Bobriki sandstones in the Ryazan-Saratov depression, as revealed by drilling operations in individual oil prospects, it was possible to prepare a very general diagram of the location and courses of the distributaries of the paleo-delta (see Figure 56). It shows that the distributaries generally had a northwest-southeast orientation, i. e., along the slope of the platform.  This means that the hydrocarbons migrated in the opposite direction, moving along numerous sand channels and accumulating either in structural or lithologic traps.  This is borne out by the variation in the bitumen content in the sediments of the coal-bearing horizon.  As one approaches the seashore, the terrigenous beds become thicker and more arenaceous; their content of bitumens also increases.

To date only oil fields of the structural type have been explored in the Volgograd region, but channel sands of the paleodelta serve as reservoirs in nearly all of the petroliferous brachyanticlinal uplifts.  These deposits, which either pinch out or are bounded by clays, are perfectly suitable for the formation of lithologically sealed or lithologically bounded oil and gas pools.  The distribution pattern of such pools associated with channel sands

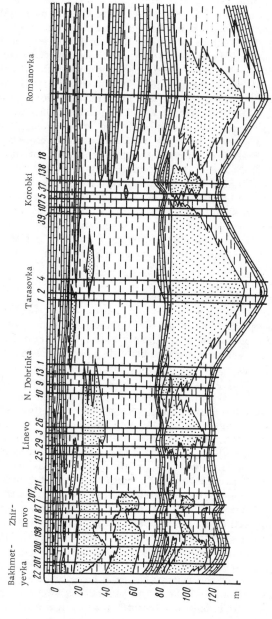

FIGURE 61.  Cross section of a part of the Ryazan-Saratov paleodelta of Yasnaya Polyana age

becomes evident if one studies the conditions of formation of known fields. The region of the Ilovlya-Medveditsa dislocations is one such example.

The hydrocarbons which migrated updip from the east passed most easily through the sand deposits of both the major paleochannel and its branches and accumulated therein. The fairly small Ilovlya uplift is one of these branches of the major migration channel. Having filled the Ilovlya trap the hydrocarbon flow moved further to the northwest in the direction of the Linevo area, filling other traps on the way. The formation of the Zhirnovo pool further to the northwest was independent of the Linevo pool, because they are separated by a strip of compact argillaceous rocks.

An oil pool discovered on the western flank of the Zhirnovo brachyanticline differs from all others in the Bobriki horizon not only in the position of its oil-water interfaces (-940 m, -911 m), but also in its reservoir pressure (lower by 23 kg/cm$^2$) and its gas factor (lower by 26 m$^3$/ton). This pool is a sand trap of the lithologic type, completely sealed by clays.

Another branch of the main channel was discovered north of the Ilovlya area, in a zone of compact sandstone. The oil here accumulated in tectonically and lithologically sealed traps situated along the path of migration of the oil. Thus, although the direction of regional migration of the hydrocarbons was the same, their economic accumulations were controlled by structural-lithologic conditions and were formed independently from one another.

Many of the delta's distributaries apparently debouched into the marine basin to the south and southeast of the main channel, as indicated, in particular, by a certain increase in thickness of the terrigenous beds in this region, as well as by the presence of oil deposits (Archeda, Zima, Saushinskoye, Verkhovye, Abramovka, etc.). These are structural-lithologic pools, many of which contain gas. The hydrocarbons originated from the same basin (the broad coastal zone of the Middle Visean sea) but migrated along different, relatively smaller channels.

It follows from the above description of the Ryazan-Saratov paleoriver and from the data cited that the territory of the Lower Volga region contains considerable accumulations of oil and gas occurring exclusively in lithologic traps. Such accumulations are particularly likely to occur in the region northwest of the Ilovlya field and east of the Zhirnovo field.

The above-mentioned areas by no means exhaust the prospects of finding new oil fields in the Volgograd region. The southeastern part of the vast deltaic zone, poorly covered by drilling because of the absence of promising structures, remains an enormous reserve for exploration of lithologic traps. The search for purely lithologic pools has yet to be started.

All major and readily established local uplifts and the associated oil and gas pools in the Volgograd region have generally been detected and explored. Therefore, the need to explore lithologic pools of fossil fuels is becoming increasingly pressing. Such exploration should primarily be conducted in zones of ancient deltas by way of detecting river beds, distributaries and channels and studying the details of their formation.

One of the most serious objections to exploration for lithologic pools is the large outlay for drilling operations. However, the use of geophysical methods has proved highly successful in recent years. These methods should be given broader application with the aim of delineating channel sands and sand bodies of other shapes. It is most expedient to conduct prospecting in areas already covered by structural, exploratory and exploitation drilling. A thorough analysis of the available geologic data and the establishment of the distribution pattern of the sand formations make it possible to locate the most suitable sites for test holes.

The above applies not only to the Volgograd region. The Volga-Ural oil- and gas-bearing region, which is the major oil-producer of the Soviet Union, requires further exploration for additional reserves. The search for new pools in this region should be based solely on paleogeographic considera- tions, since nearly all of the known structural traps have already been drilled. It is necessary to carry out detailed paleogeographic reconstructions of the Devonian, Lower and Middle Carboniferous and Lower Permian producing formations. This will enable us to pinpoint the most promising areas for exploration of lithologic and stratigraphic pools which remain the only re- serves for maintaining production at the present level for a reasonably long period of time.

## 14.  STRUCTURAL AND PALEOGEOGRAPHIC ASPECTS OF THE GENERAL DISTRIBUTION PATTERN OF OIL AND GAS

Empirically it has long been noted that oil and gas pools most often occur in the peripheral zones of regional depressions or basins. Most researchers apply the tectonic principle, i.e., the present-day structural pattern, in order to identify territories of regional oil and gas accumulation. However, the process of formation and the distribution pattern of oil and gas fields are often studied without taking into account their evolution, i.e., disregarding the paleotectonic pattern. It would obviously be more correct to link oil- and gas-bearing territories not only with present-day depressions but also with ancient ones. The latter have undergone repeated changes both in time and space.

Why are oil- and gas-bearing territories associated with depressions rather than with uplifts? Is this not at variance with the fact that oil and gas pools occur in anticlinal structures?

Firstly, depressions, which serve as sedimentary basins, are regions of accumulation of both source beds and reservoir rocks. We have already pointed out the tremendous role played by bacteria which form part of the biomass of shallow seas. We wish to emphasize the importance of shallow- water environments for the proliferation of microorganisms. The largest concentration of organic remains, which are the starting material for the formation of bitumens, usually occurs in the coastal zone of marine basins. This is why source beds are generally restricted to the marginal portions of regional depressions.

Secondly, structural oil and gas traps are usually local uplifts within re- gional depressions. The following example may be cited in order to support this statement: according to Bentz (1961), oil in the Northeast European basin is produced mainly from Jurassic and Lower Cretaceous sediments. Only 10—15% of the 200 salt domes contain oil.

Bentz attributes the absence of oil in the majority of salt domes to the fact that they are situated beyond the limits of the main Mesozoic depressions.  Vast areas lacking oil are associated with ancient uplifts which experienced sedimentation on a much smaller scale during the Jurassic and Lower Cretaceous than did the depressions.  The latter are the regions where source beds were formed.  Traps of various types, within which the migrating oil accumulated, occur mainly in the marginal portions of the depressions.

Is the tectonic or even paleotectonic criterion, taken on its own, a sufficient basis for assessing the oil and gas prospects of a particular territory? When one considers all of the various factors involved in the complex process of formation of oil and gas pools, such prerequisites as the presence of both source beds and reservoir rocks are no less important than tectonics.  Firstly, the source beds must have a relatively high productivity; secondly, the reservoirs must possess a porosity and permeability sufficient for the accumulation of hydrocarbons in structural, lithologic or stratigraphic traps.  In our opinion, the thickness of the reservoir rocks is a decisive factor in the formation of rich deposits, since the high productivity of the source beds may be due either to the vast territory of their occurrence or to the prolonged existence of conditions favorable for normal transformation of the organic matter into hydrocarbons.

None of these factors, when considered on its own, can be regarded as sufficient for the formation of a gas or oil pool.  The complete process of formation takes place only if there is a certain combination of all the necessary factors.  However, the above comment does not rule out the need to study each of these factors separately in order to gain a better understanding of the process as a whole, the more so since some of them appear to be dependent on several conditions or other factors.  One of the latter is the physiographic environment of accumulation of the producing formation, which is in turn a combination of individual factors such as tectonics, relief, lithofacies conditions, climate, etc.  We therefore believe that the paleogeographic factor, which includes a number of major conditions of sedimentogenesis, ranks among the most important ones and may have a decisive influence on the formation and accumulation of hydrocarbons in certain traps.  This is quite evident from the example of the paleogeographic conditions of formation of the producing horizons of the Yasnaya Polyana substage, as well as of many other strata containing oil and gas.

We believe that one of the most important criteria for the presence of oil and gas within a certain lithostratigraphic complex is its position relative to the coastal zone of a marine basin during the period of accumulation of the oil-bearing beds.  An objective assessment of the prospects of such a complex is to a considerable degree dependent on the detail of the paleogeographic reconstruction.

The practical experience gained during worldwide oil production and the data presented in this book show convincingly that economic concentrations of oil and gas are generally confined to shorelines of marine basins possessing gently sloping shelves.

We have already mentioned the association of oil fields with shorelines in many parts of the globe.  Russell (1958) pointed out that the oil fields on

the Gulf Coast and in Texas, Louisiana and other regions of the U. S. A.  coincide almost completely with the ancient shorelines for each producing
formation.

The association of oil pools with seashores is also confirmed by the fact
that in recent years the greatest strides have been made in oil exploration
and production in shallow-water areas of the continental shelf.

Rich oil and gas areas are situated along the coast of South America: the
Maracaibo and Eastern Venezuela basins.  The latter is the vast deltaic region of the Orinoko River which has been flowing into the Gulf of Venezuela
for a very long period of time.

In Europe, the shoals of the Mediterranean basin, remains of the ancient
Tethys, are promising areas for oil and gas.  In particular, the rich oil-
and gas-bearing region in the lower reaches of the Po valley is adjacent to
the shallow northwestern part of the Adriatic Sea.  Large French oil pools
on the coast of the Bay of Biscay are also associated with deltas of major
rivers.  For instance, the very rich Parentes deposit near Bordeaux is in
the delta of the Garonne River.

Unique oil fields are known to occur on the coast of the relatively shallow
Persian Gulf where offshore areas are presently also being explored.

Also rather promising are the shelf zones sloping gently into the seas and
oceans off the coasts of Indonesia, Japan, Burma, Australia, North Africa and
elsewhere.  We are convinced that on the shores of the numerous seas fringing the Soviet Union large accumulations of oil and gas await discovery.

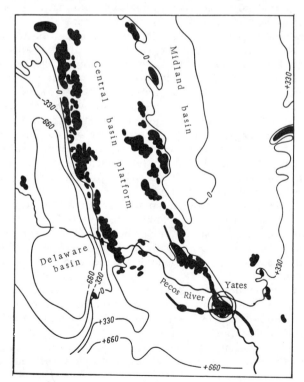

FIGURE 62.  Location of oil and gas pools in the Permian basin, U.S.A. (Ver Wiebe, 1952)

The zonality of oil and gas accumulation, linked with both tectonic and paleogeographic factors, is well expressed, for instance, in the Permian basin (U. S. A.), where the producing horizons are arenaceous and carbonate rocks ranging in age from Ordovician to Permian.   Oil and gas fields here are arranged in belts extending for tens or even hundreds of kilometers (Figure 62).   The western belt consists of littoral terrigenous-anhydrite beds which give way along the strike to reef formations which contain the largest oil and gas deposits.   This belt is about 150 km long and 5—20 km wide.

Several hundreds of deposits have been discovered in the Permian basin. From the tectonic viewpoint most of them are situated on the slopes of the Central basin platform, while from the paleogeographic viewpoint they are located along the ancient seashore.   More than a half of the entire production of this region is from 12 major fields.

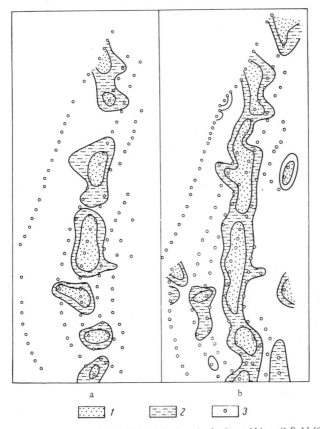

a                                      b

☐···☐ 1          ☐‒‒☐ 2          ☐ o ☐ 3

FIGURE 63.   Distribution of sand reservoirs of the $D_1$ horizon in the Romashkino oil field (Cholovskii, 1966): a — as of 1 June 1959; b — as of 1 June 1963; 1 — high-conductivity reservoirs (sandstone); 2 — low-conductivity reservoirs (siltstone); 3 — wells.

The distribution of sand reservoirs and, consequently, the accumulation of oil and gas in local uplifts are frequently controlled by the physiographic environment of sedimentation in the past.  For instance, the bulk of the oil in one of the largest fields in the Soviet Union — Romashkino — is located in the well-sorted sandstones of the Pashski horizon.  The distribution pattern of these sandstones is obviously prodeltaic.  Individual strips of these channel sands range in width from 0.5—1.5 to 6—8 km.  They form the $D_1$ horizon and have a very complex structure, being made up of several interlayers or lenticular bodies.  Their overall thickness ranges from a few meters to 15—30 or even 50 m.

Locally, elongated "patches" of argillaceous siltstone or mudstone occur among the channel sands.  One can assess the structure of one of the producing horizons from Figure 63 which shows how many dry holes were drilled because the genetic features of the sandstone distribution were not taken into account.

Levorsen (1958) noted that oil and gas zones occur on the margins of large depressions.  He emphasizes stratigraphic as well as structural traps. I. M. Gubkin also admitted in fact that zonality exists when he spoke of the influx of oil and gas into anticlinal zones from the oil-collecting areas in the adjacent depressions.  Brod (1946) also wrote in this sense when distinguishing oil and gas accumulation zones which are inseparably linked with zones of oil and gas generation.  The latter zones were classified by Brod as elements of large depressions in the present-day structure of the earth's crust.

Most investigators attribute the zonal distribution of oil and gas pools to the tectonic structure of the crust.  This, however, is not the full picture. In addition to its dependence on the character of regional depressions, zonality is also controlled by the paleogeographic environment.  The latter helps us to solve very important problems: establishing the source from which the oil and gas migrated and the sites where they could accumulate to form commercial deposits.

*Distribution Pattern of Large Hydrocarbon Accumulations*

In addition to the general distribution pattern of oil and gas, the particular patterns of the large deposits are also of great practical significance.  We have already established that the areas of maximum concentration of these fossil fuels are very often associated either with regions of ancient deltaic formations or with alluvial-fluviatile sediments of coastal zones, or with reefs.

The proved and estimated oil reserves of the Volga-Ural region and the Dnieper-Donets depression, as well as of other oil- and gas-bearing regions, both in the Soviet Union and in other countries, provide rather convincing evidence of this.

The Timan-Pechora and, partly, the Kizel paleodeltas are somewhat exceptional among the delta zones in the Ural-Volga region, in that their sediments are to a considerable extent incorporated into the folded zone of the Urals and exposed at the surface.  However, here too the former existence of extremely rich deposits of oil and gas in the Timan-Pechora delta is

indicated by the thick "grindstone" suite which is in fact a quartzose sandstone cemented by oxidized oil whose content exceeds 4% of the overall
weight of the rock.

The mouth of the Kizel paleoriver and especially its prodelta were also
incorporated into the folded region of the western slope of the Urals by the
Hercynian orogeny and have undergone considerable erosion. However,
under these conditions too the richest oil fields of the Ural region near
Perm are located in the estuarine part of the ancient river valley. Among
the largest oil fields discovered to date are those in Yara-Kamenny Log
region.

For a more complete assessment of the productivity of the oil fields
associated with paleodelta formations it is preferable to concentrate on
those not affected by erosion. For example, in the Mukhanovo field, located
in the central part of the Early Visean paleodelta, 60% of the total reserves
occurs in the Lower Carboniferous and the remainder in the Devonian.

The Kuleshovo, Mikhailovka-Kokhanovo and Dmitriyevka oil fields in the
same deltaic zone also rank among the richest.

The reserves of the Radayevka field further up the river valley of the
paleo-Kama are much poorer than those concentrated in the sand deposits
of the ancient delta. No oil whatsoever was produced from the Melekess
test hole drilled even higher up the valley of the same paleoriver.

A characteristic feature of the rich natural reservoirs associated with
delta and estuary formations becomes even more evident when one compares
them with other areas lacking alluvial-fluviatile deposits. For example,
within the Tatar Arch, the average thickness of the Yasnaya Polyana substage
is relatively small and its sediments, represented by continental facies remote from the sea, virtually do not contain any oil pools. The only exception is the Novoyelkhovo field where only a small portion of the reserves is
contained in the Bobriki horizon. It is, however, not known whether this oil
is of the same age as the enclosing rocks.

The Bobriki and Tula horizons in the Saratov region are productive, but
their oil reserves are inferior to those in other fields of Lower Carboniferous age. In the Sokolovgrad field the Bobriki horizon contains little
more than 2% of the total reserves of the producing formations, including
rocks of the Devonian System. Such a proportion in most fields of the
Saratov region points to the minor significance of oil in the Carboniferous
deposits.

We believe that one of the main reasons for the relatively small oil accumulations in the right-bank areas of the Saratov region is their great distance from the sea and the lack of sufficiently thick alluvial deposits. However, in the Transvolga region closer to the ancient seashore, where beach
and bar sandstones were deposited, larger pools may be encountered.

The situation is completely different in the estuarine part of the broad
valley of the Ryazan-Saratov paleoriver. Much larger oil accumulations
are encountered in this zone. The greater part of the proved oil reserves
in the Korobki field occurs in the Bobriki horizon. In the Zhirnovo field
the major reserves are in the Bobriki and Tula horizons, while in the
Bakhmetyevka field they occur within the Yasnaya Polyana substage.

In addition to deltaic sediments, other alluvial formations deposited near the shores of ancient seas also served as suitable reservoirs for migrating hydrocarbons. A good example are the reserves in the oil deposits in the Birsk saddle of Bashkiria. The well-known Arlan field, situated in the zone of a river paleodelta extending for quite a distance along the marine basin, contains large reserves of oil confined mainly to alluvial-fluviatile deposits of Middle Visean age.

Oil pools located in deltaic deposits of a Middle Visean river were also discovered in the Shkapovo and other regions in the southeast of Bashkiria where the oil reserves in the Lower Carboniferous sandstones are rather large.

Appreciable reserves of oil are contained in the Gnedintsevo, Kachanovka, and Glinsk-Rozbyshevo fields discovered in the Carboniferous deposits of the Dnieper-Donets depression. They are situated within the delta region of the Pripyat paleoriver.

We could cite quite a few other examples corroborating the association of the large oil pools with similar deltaic zones of the East European platform. It will suffice to recall the known Romashkino field in which the greater part of the oil occurs in the Devonian deposits represented by typical prodelta sediments.

The richest oil fields in the U.S.A. were discovered on the Gulf Coast where they occur in thick deltaic formations of the Mississippi. According to the latest estimates, the coastal strip of the Gulf, within the state of Louisiana and constituting only part of the delta, contains one-third of the total reserves of the U.S.A.

These data support the opinion that the distribution of large oil and gas fields follows a certain pattern. However, detailed paleogeographic studies are necessary before one can make use of this regularity for exploration purposes.

The map in Figure 64 compares oil reserves in deposits situated in the paleodelta zone with those discovered in other areas, according to individual economic regions. It follows from the map that the largest quantities of oil in the Lower Carboniferous sediments occur in estuarine formations of ancient rivers, thus confirming the conclusion that deltaic formations play a special role in the accumulation of large oil deposits.

There is abundant factual evidence indicating that such a distribution pattern for large oil pools is not distinctive only of the Lower Carboniferous producing formation on the East European Platform. This is why it is so important to make use of this regularity in order to increase the efficacy of oil exploration. In this context, the West Siberian Lowlands may be regarded as a vast territory meriting investigation.

As far back as 1932, I. M. Gubkin suggested that the Jurassic coal-bearing deposits of West Siberia, distributed along the eastern slope of the Urals, show transition to source beds. "If one assumes that the Jurassic shore facies — the facies of lakes, swamps, limans and lagoons — was the site of accumulation of the organic material from which the coal was later formed, then the nearshore facies of that same Jurassic sea could also have been a suitable site for the accumulation of sapropel material which could have served as the source for the formation of oil. One can thus expect a transition from coal facies to petroleum facies somewhat further to the east. Oil prospecting in the Jurassic on the eastern slope of the Urals is, therefore, wholly justified."

Gubkin's assumptions have been confirmed. Such a change from obviously continental coal-bearing formations to nearshore-marine oil- and gasbearing rocks of both Jurassic and younger age is noted as one moves from the periphery to the center of the West Siberian Lowlands and is also characteristic of its individual first-order negative structures.

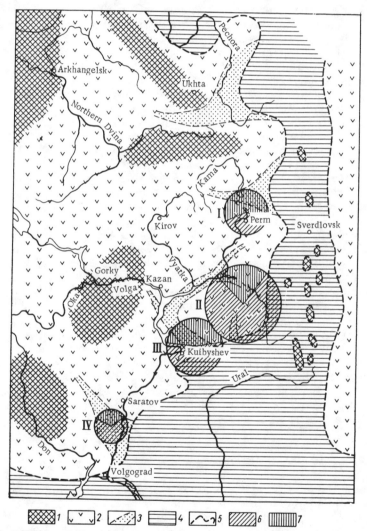

FIGURE 64. Comparison of oil reserves in the Yasnaya Polyana substage in various paleogeographic zones of the Ural-Volga region:

1 — elevated portions of the paleorelief; 2 — lowland plain; 3 — valleys and deltas of paleorivers; 4 — marine basin; 5 — shoreline; 6 — reserves occurring in areas of deltaic and alluvial-fluviatile deposits; 7 — reserves on the rest of the territory. I — Perm; II — Arlan; III — Kuibyshev; IV — Volgograd.

A very general comparison of the structure of the East European Platform with that of the West Siberian Lowlands or plate reveals that while
the former was essentially an enormous monocline, the latter, fringed by
uplifted regions (the Urals, Northern Kazakhstan, Altai-Sayan Range, Yenisei
Range, and Taimyr), was generally downwarped in its central regions.   The
huge West Siberian sedimentary basin had a complex tectonic structure,
and its most downwarped portions had long been occupied by marine basins.
Near the shores of the latter the environments were favorable for the formation and accumulation of hydrocarbons.

Other investigators have also attempted to reveal the distribution patterns
of large oil fields.   Maksimov (1964), for instance, associates the formation
of such fields with the following five regularities:

1) correlation with tectonic zones — swells — which complicate depressions and adjacent slopes of arched uplifts in zones of increased thickness
of terrigenous complexes;

2) association with swells, tectonic zones and structures complicating
arches within which the thickness of the terrigenous complex is not great,
but where the structural elements mentioned are situated near depressions
in which the terrigenous deposits attain maximum thickness;

3) accumulation of hydrocarbons according to the principle of differential
entrapment;

4) inherited growth of fairly large local uplifts or even larger structural
elements providing favorable conditions for the formation of highly productive pools in the relatively elevated parts of the tectonic zones, owing to
regional lateral migration and, especially, re-formation of other pools;

5) occurrence of the lightest oil (gas) in relatively deep traps situated in
zones where the terrigenous rocks attain maximum thickness.

Maksimov himself admits that the last-mentioned regularity is by no
means universal.   Moreover, some of the other regularities are contra-
dictory and therefore a combination of two or three of them is required for
the formation of highly productive fields.

Such conclusions, based on actual observations, are generally empirical
and do not explain the causes of the above-discussed regularities.   They
may be elucidated only by reconstruction of the paleotectonic and paleogeographic environments in which individual productive lithostratigraphic
complexes were formed.

Consequently, paleogeographic reconstructions, complementing tectonic
analyses, are essential in order to establish the distribution patterns of
large and small oil and gas deposits.

*Belts and Major Zones of Oil and Gas Accumulation*

A study of the physiographic conditions of deposition of producing formations and the distribution of the oil and gas fields within them (Figure 59)
leads to a definite concept of the zonal distribution pattern of oil and gas
within the paleogeographic scheme.   This concept has been confirmed by
practice which is always the best check on theoretical assumptions.

Coal geologists are familiar with the "theory of belts and major zones
of coal accumulation" advanced by Academician P. I. Stepanov (1947).   The

gist of the theory is that the accumulation of coal throughout the globe is subject to well-defined regularities and is the result of an entire complex of geological phenomena.

Those areas of the earth's surface in which particularly large coal-bearing deposits and coal bodies accumulated during a specific geologic period were termed "belts of coal accumulation" by Stepanov. Within these belts he distinguished smaller areas with especially numerous coal basins which he termed "major zones of coal accumulation."

Stepanov's identification of belts and major zones was based on an analysis of known coal reserves on a global scale. However, he admitted that his theory was incomplete. In particular, he wrote: "It would be more appropriate to view this theory as the first stage of research — a 'hypothesis' on belts and major zones of coal accumulation. Vernadskii, in private talks on belts and major zones of coal, recommended that this theory, at its present stage of development, be regarded as 'empirical laws governing the process of coal accumulation.' Vernadskii proved to be right. Until all the factors responsible for the formation of belts and major zones are analyzed comprehensively, my work cannot be considered complete."

This shows how great a significance Stepanov attached to discerning the causes of the distribution patterns of coal. We are of the opinion that paleogeographic studies reveal these causes to a certain extent.

Similar ideas on the zonal distribution of oil pools were later adopted in petroleum geology. Khain (1954) distinguishes belts of bitumen accumulation which contain several closely spaced oil and gas basins having a common trend and similar geotectonic features, the main oil-bearing formations being of the same age. Belts of bitumen accumulation are particularly prominent in elongated geosynclinal belts. In platform regions these belts coincide with zones of maximum subsidence, which extend parallel to the most active geosynclines of the corresponding time period.

Khain tends to classify areas where the principal belts are intersected by transverse depressions as major zones of bitumen accumulation. The belts become wider in such areas, centers of steady subsidence develop there, and oil and gas occur over a wide stratigraphic interval. Khain divides the major zones into small and large. The large major zones are regions of intersection or merging of large belts of the same or different age. Examples of such major zones containing maximum oil reserves are the region of the Caspian Sea and Persian Gulf in the Eastern Hemisphere and the Caribbean and the Gulf of Mexico in the Western Hemisphere. These regions are referred to as the "oil poles of the world."

Brod (1960) classified large regions of crustal subsidence forming structurally uniform groups as oil- and gas-bearing belts. He distinguishes five geotectonic belts containing oil and gas basins. Three of these belts contain basins associated with the largest mountain structures, while the other two contain basins situated on platforms (mainly plain areas). The basins of two of the mountain belts fringe the Pacific Ocean on the west and east, while in the third the basins form a broad band extending latitudinally across the Eastern Hemisphere.

The two platform belts are classified as northern and southern. The former includes the oil and gas basins on the North American Platform, the slopes of the Baltic Shield, the nonalpine parts of Western Europe, and the East European and Siberian platforms. The latter includes basins located in large regions of subsidence on the Brazilian, African and Australian platforms. According to Brod, the grouping of oil and gas basins as elements of geotectonic belts is the first attempt at revealing their distribution pattern. Other investigators also subdivide the territories of regional oil and gas accumulation into belts or zones.

For example, Bakirov (1959) also includes in the category of belts of oil and gas accumulation smaller units, such as oil and gas provinces and regions situated within a certain fold system and connected genetically with the formation and development of a given geosynclinal belt. These belts are classified as geosynclinal and transitional territories which are piedmont basins and foredeeps of platforms.

Other oil-bearing territories are classified according to tectonic and stratigraphic criteria, or to the chemical type of the oil (V. A. Uspenskii, O. A. Radchenko and others). However, we yet lack a hard and fast rule for classifying large oil- and gas-bearing territories, and most scientists resort to the geotectonic principle when identifying belts and major zones of oil and gas accumulation.

It can readily be seen that the identification of such belts on a global scale is insufficient for the purposes of oil and gas exploration. The belts, and especially the major zones, must serve as a starting point for the choice of the optimal course of exploration work. The more detailed the knowledge of the structure and location of such belts and zones, the better substantiated and effective will be the search. This becomes possible only if all the factors responsible for the zonal distribution of economic accumulations of hydrocarbons are taken into account. Thus, the tectonic principle employed in the identification of belts and major zones should definitely be supplemented by data on the paleogeologic environment in which oil- and gas-bearing deposits were formed.

The above analysis of the distribution of oil and gas pools in a number of the world's oil and gas regions indicates the advantage of identifying belts and major zones of oil and gas accumulation on a paleogeographic basis. The erosion of rocks and the transport of debris and their transformation into sediments, as well as the saturation of the latter with organic matter, are all controlled by the physiographic environment (the joint product of climate and tectogenesis), the extent of development of organic life, the bio- and geochemical medium, and other factors. If the geotectonic regime affects sedimentation it is via the physiographic environment during certain periods of the geologic history of a particular territory. It is, therefore, good practice to take into account the tectonic, paleotectonic and paleogeographic factors when studying oil- and gas-bearing terranes.

Reconstruction of a past paleogeographic environment does not merely involve reconstruction of the ancient landscape, but also calls for determination of the qualitative and quantitative changes in the biosphere in time and space. The biosphere plays a decisive role in the formation of hydrocarbons. The conditions most conducive to the development of organic life usually arise in relatively shallow-water zones of marine basins exposed to sunlight, on continental shelves, as well as on lowland coasts with numerous water bodies.

The coastal zones of seas and oceans are the sites not only of maximum accumulation of primary organic matter for the formation of hydrocarbons, but also of intensive sedimentation. This combination explains why conditions suitable for the formation of source beds and reservoir rocks frequently arise in such zones. The absolute majority of the world's largest oil and gas fields are located along the coasts of ancient and present-day seas. An example of the dependence of oil and gas accumulation on the physiographic factors is provided by the shelf zones of present-day seas, from which more than 17% of the total oil production of the developing and capitalist countries is obtained. New water areas are being explored every year. Exploration drilling is being conducted along the coasts of many European countries, as well as in Tunisia, Lybia, Saudi Arabia, Iran, West Africa, Pakistan, India, Australia, Alaska and North and South America.

Along the west and east coasts of Canada geophysical surveys are being conducted over an area of 554,000 km$^2$. Oil production from marine areas of Southern California amounts to 570,000,000 tons; the major oil fields on the adjacent land areas (Huntington Beach, Seal Beach, Long Beach, Wilmington, etc.) have already yielded more than 700,000,000 tons of oil. The presence of such rich deposits on the east coast of the U.S.A. is probably due to the fact that nearshore-marine conditions existed from Alaska to California during the Paleozoic and Early Mesozoic.

In Central America the most promising oil and gas prospects are the territories extending along the coasts of the Pacific and Atlantic oceans. Many companies are engaged in oil exploration in Iran, in the waters of the Persian Gulf and in the continental area. In 1965 the first Alaskan oil was produced from the present-day shelf zone. Five marine fields have already been discovered in Cook Inlet.

According to L. Weeks (1966), the largest Mesozoic-Cenozoic oil basins are situated close to the coasts of present-day seas and fairly frequently in the seas themselves. Although such statements are quite common in the geologic literature nowadays, convincing explanations of this phenomenon are not as yet forthcoming.

The number of countries engaged in marine oil exploration has increased eleven-fold over the past decade, and now totals 65. The proved reserves of "marine" fields account for nearly 18% of the global oil reserves. According to Gardner's (1966) estimate, marine oil reserves amount to 96 billion tons. The potential oil resources on the continental shelf of Canada alone are estimated at 8.2 billion tons, and those in the oil basin of Western Canada, at 4.8 billion tons.

The Beaufort continental shelf and the adjacent Arctic coastal plain hold great promise. Of special interest is the delta of the Mackenzie River where drilling was recently started. Along the Pacific coast of Canada, the shelf, consisting of Cenozoic deposits, possesses all the features of a fairly promising oil- and gas-bearing territory. In particular, in the Hecate region the continental facies give way to marine ones. The lagoonal part of the Hecate Strait is a possible zone of oil and gas accumulation. American geologists believe that the west coast of Canada and the Californian coast of the U.S.A. have similar structures.

Marine oil exploration has been started off the Atlantic coast in Brazil. One of the most promising localities is the Reconcave offshore region where an oil field was discovered, part of which is situated on the continent.

Chile and Peru were the first to achieve notable success in the sea. Offshore drilling is being carried out in Nicaragua, Guatemala and British Honduras.

Indonesia is an important oil-producing country whose numerous fields are located in the deltas of Cenozoic and younger rivers. Offshore areas in the vicinity of some of the islands hold great promise.

In Australia the thick Cenozoic deposits within the territorial waters are the main object of exploration. Offshore drilling is being conducted in the Bass Strait and in Gippsland (coastal waters of Victoria). Several oil and gas fields have already been discovered.

Large gas fields were discovered in The Netherlands (Groningen Province), calling for a reappraisal of the gas prospects in the North Sea. Highly porous Permian sandstones in the Groningen Province, up to several hundreds of meters thick, were found to be gas reservoirs. It was later established that the gas-bearing formations in the coastal regions of The Netherlands, West and East Germany and Great Britain extend into the sea. Large-scale exploration for oil and gas in the North Sea is now underway.

The above examples vividly illustrate the zonal distribution of oil and gas fields along the shores of ancient and present-day seas. We have already discussed the reasons for this association. Since seashores are usually by nature striplike, the oil- and gas-bearing zones extending along them can be considered as belts of oil and gas accumulation. But this is not all. Our observations have shown that the largest oil fields are associated either with the deltaic deposits proper, the adjacent sand-silt formations, or reef bodies.

The significance of estuarine sediments as natural reservoirs of maximum oil accumulation has been discussed earlier in this book. We shall only recall that the richest oil regions of the Soviet Union (Romashkino, Apsheron, Mangyshlak, Shirotnaya Ob region, Kinel-Cherkassy, etc.) are associated with zones of deltaic or near-delta formations built by ancient rivers. The world's largest oil fields are located near the mouths of rivers, such as the Mississippi, Orinoco, Shatt al-Arab (the mouth of the Tigris and Euphrates), Niger, Irrawaddy, Mackenzie, Nile, Garonne, Po, etc. There is no doubt that large oil pools will soon be discovered in the estuarine zones of several large ancient rivers.

Recent practical experience as well as our own research provide incontrovertible evidence of the existence of a certain regularity in the distribution of large oil fields within belts of oil and gas accumulation. This makes it possible to distinguish major zones of accumulation of liquid and gaseous hydrocarbons. Figure 59 shows one such example.

Thus, we classify as belts the broad and elongated zones on both sides of the land-sea boundary line, and as major zones of oil and gas accumulation — areas of estuarine deposits where rivers debouched under conditions of relatively steady subsidence. It is precisely in these localities and also on the enormous adjacent territory of the sea bottom that organic matter, together with mineral material, accumulated and was subsequently transformed. In some cases the mineral material played the role of a catalyst in the process of transformation of the organic matter into hydrocarbon compounds, while in others it served as an accumulator of the final products.

Recognition of belts and major zones of oil and gas accumulation according to paleogeographic criteria makes it possible to re-evaluate oil and gas prospects and to select the most rational trends of exploration. In view of the possibility of forecasting and organizing exploration of large fields, the identification of major zones of oil and gas accumulation takes on special importance.

## 15. CERTAIN FEATURES OF OIL AND GAS ACCUMULATION IN WEST SIBERIA

The West Siberian oil and gas megabasin ranks among the world's largest sedimentary basins. One of its characteristic features is the absence of young, vigorously developing geosynclines along its margins; its territory lacks exogeosynclines or foredeeps. The basin has broad and relatively gently sloping southern and southeastern flanks and steeper and narrower western and, partly, eastern flanks. Toward the north the basin links up with the Arctic Ocean. Individual positive and negative elements of various orders were formed against the general background of regional subsidence.

We shall not present a detailed description of the structure of the West Siberian basin, as such has already been provided by N. N. Rostovtsev, V. D. Nalivkin, M. Ya. Rudkevich, G. I. Boyarskikh, F. G. Gurari and other investigators. We only wish to remark that its evolution, as well as the changes in the physiographic sedimentary environment, was determined by tectogenesis. The course of events during the Mesozoic-Cenozoic is roughly as follows.

During the Early and Middle Jurassic the greater part of West Siberia was a humid lowland covered by forests, with numerous swamps, lakes and rivers. A marine basin was located to the north of this lowland, individual gulfs of which occasionally encroached southward. The end of the Middle Jurassic witnessed the onset of a marine transgression which attained its maximum in Late Jurassic—Early Valanginian times.

A regression took place during the Hauterivian—Barremian, resulting in a certain decrease in the area covered by the sea. However, a new transgression commenced in the Late Barremian and continued throughout the Albian and Aptian. Another major regression occurred during the Cenomanian, the seashores shifting toward the north and center of the basin. Next there followed a more protracted transgression (Santonian-Campanian-Maestrichtian). After a minor regression during the Paleocene, a transgressive marine regime was re-established in the Early and Middle Eocene.

As the transgressive and regressive cycles of sedimentation superseded one another, the position of the seas and their shores and of the paleorivers and their mouths also shifted, the depositional environment underwent transformations, and favorable or less favorable combinations of reservoir rocks, source beds and cap rocks arose. Anticipating the later discussion, we feel obliged to mention here that, when one compares the distribution of discovered oil and gas fields with the paleogeographic environments at the time when the major producing formations accumulated, one notes that the vast majority of the fields tend to occur in ancient coastal zones. This is indicative, firstly, of the regular zonality of the oil and gas accumulations and, secondly, of the leading role played by the paleogeographic factor in the distribution of oil and gas pools.

The producing formations of the West Siberian oil and gas basin are exclusively represented by Jurassic and Cretaceous sand and sand-silt deposits. The results of exploration and research have made it possible to identify up to seven independent oil- and gas-bearing strata. They are described briefly below.

The Lower-Middle Jurassic section is made up of a terrigenous complex deposited mainly in lacustrine, lacustrine-palustrine, alluvial and proluvial environments. Its thickness is 300—500 m in depressions and only tens of meters (sometimes even totally missing) in uplifted areas.

Sandstones account for 30—55% of the entire section, the thickest beds occurring in the central part of the basin. The proportion of sandstones increases to 80—90% in the southern and southeastern parts. For instance, 29 sand beds were found in the Ob-Irtysh interfluve among deposits of the Tyumen suite. It has been ascertained that most of these sand beds are of fluvial origin. The sandstones usually occur as lenticular bodies of various size. Not infrequently they give way to silty and argillaceous facies over small distances.

The Upper Jurassic formations in the western part of the region are marine sediments, the uppermost layers being bituminous mudstones. As one approaches the pinch-out zone of the Lower Jurassic rocks, the section contains more sandstone, and the bitumen content of the mudstones virtually drops to zero. The lower part of the series in the eastern regions features nearshore-marine sandstones which become more widespread toward the margins of the basin. Mottled clay deposits of the lagoonal type make their appearance in the southern regions. They are often underlain by basal sandstones showing variable distribution and inconsiderable thickness (1—15 m).

On the whole, the thickness of the Upper Jurassic deposits varies from 40—50 to 600—700 m within the West Siberian basin. It attains a maximum (recorded in the Ust-Yenisei depression) along the northeastern margins of the basin. The combined thickness of the sand-silt sediments accounts for 5—10 to 60% of the section.

The Lower Valanginian formations are represented by arenaceous-argillaceous and argillaceous rocks of marine origin. The arenaceous varieties have a fairly wide distribution, the only exception being the western regions. In the eastern and central regions of Tyumen the arenaceous rocks make up the Achimovski series. Irregular distribution is a characteristic feature of these rocks. They generally occur as lenticular bodies several kilometers long and up to 10 m thick; locally their thickness attains 25—30 m. In some places the sand lenses are separated from one another by argillaceous rocks, while in others they link up to form complex systems. The thickness of the Achimovski series ranges from 5—10 to 50—70 m.

The Upper Valanginian consists mainly of argillaceous rocks; arenaceous rocks occur in the eastern region of Tyumen. In the southern and central regions sandstones frequently alternate with mudstones.  West of the Nizh-nevartovsky oil field the clay content increases abruptly, and further west the series becomes almost exclusively argillaceous.  An increase in the sand content has been noted in the Surgut arch, especially in its southern part where a group of productive beds, consisting of fine-grained, in places silty sandstones, makes its appearance.  The thickness of the Upper Valanginian series varies from 10—20 m on the margins of the lowland to 100—−150 m in the central region, increasing to 300—500 m in the northern region. The combined thickness of the sandstones varies from 5—10% in the zone where argillaceous rocks predominate to 80% in areas adjacent to the Tai-myr and Altai-Sayan uplifts.  More than 30 oil fields have been discovered to date in the Upper Valanginian deposits, mainly in the middle reaches of the Ob River.

The Hauterivian-Barremian formations are composed of nearshore-marine and continental sandstones, siltstones and clays.  In the southern regions mottled arenaceous-argillaceous formations of the Kiyaly series are widespread.  To the north they are replaced by green varieties of the Vartovsky series.  In the central part and on the western flanks of the Khan-ty-Mansi and Nadym depressions the Hauterivian-Barremian deposits con-sist of argillaceous rocks.  In the western regions individual arenaceous members are encountered among nearshore-marine and marine clays.  The thickness of this rock complex increases from the margins of the basin (100—150 m) toward its center (500—900 m).  The maximum thickness was recorded in the Ust-Yenisei depression.

Sandstones account for 5—10% of the section in the central regions and for 60—80% in the eastern regions.  The highest permeability was noted for the sandstones developed on the margins of the lowland, i. e., in areas close to the provenance.  Their effective porosity here reaches 25—30%, and their permeability is 1 darcy.  The reservoir properties of the sandstones be-come somewhat less favorable toward the center of the lowland.  The sand-stones with good reservoir properties usually form discontinuous pools, generally having channel or lenticular shape.  For instance, in the Vartovsky series the lenticular sandstone formations give way to argillaceous rocks over short distances or combine to form an intricate, hydraulically linked system.

In the southern part of the lowland, the content of sand-silt rocks in the Hauterivian-Barremian section increases toward the east.  The combined thickness of these rocks increases from 50 m in the Salym region to 140—−180 m on the Surgut Arch and attains 180—380 m within the Megion and Vakh uplifts.  The reservoir properties of the arenaceous rocks improve in the same direction.

The Aptian-Albian-Cenomanian section consists of alternating sandstones, siltstones and clays.  Over the greater part of the territory of their occur-rence continental and nearshore-marine facies are common.  In the south they give way to mottled lagoonal or deltaic facies.  The thickness of the sec-tion shows a general increase from the periphery (100—200 m) to the center of the basin (800—1000 m).

Fairly large gas fields were discovered in arenaceous rocks of the Ap-tian-Albian-Cenomanian formations: Urenga, Purneiskoye, Taz, Komsomolsky, Zapolyarnoye, etc.  Several oil pools have also been discovered.

The Upper Cretaceous (excluding the Cenomanian) is made up of marine clays and siliceous-argillaceous rocks. The quantity of arenaceous rock types in the section increases in the eastern and southern regions. The total thickness of the section increases from 50–100 m along the margins of the lowland to 300–500 m in its central part and to 700–800 m in the north (Taz bay and other areas). The combined thickness of the sandstones accounts for 5–10% of the section in the center and 50–60% at the margins. Sand deposits are especially widespread in eastern regions. The Severnoye, Zapolyarnoye and other gas fields were discovered in the Upper Cretaceous sequence, while in the Taz area oil was found in Turonian deposits.

The following questions arise. Why, in similar structural environments, are oil and gas fields discovered one after another in some regions, whereas in others they are absent? Why are there so few productive fields in the southern and southeastern regions (the Omsk-Tara region near Irtysh, the Verkhne-Demyanovka megabar, the Paidugina bar, etc.)? These questions yet await answers.

We believe that the clue to the solution of this complex problem lies in detailed paleogeographic reconstructions, naturally, in combination with other studies. In fact, it would be unreasonable to expect the occurrence of oil in the purely terrestrial Lower and Middle Jurassic deposits in the southern regions. The character of these deposits is entirely different in the north, where a sedimentary environment transitional from continental to nearshore-marine had existed for a relatively long interval of time. Here one can reliably assume the presence of an independent zone of oil and gas accumulation along the shore of the Early and Middle Jurassic sea.

The Late Jurassic sea with its gulfs, bays and lagoons encroached far to the south and southeast of West Siberia. Its shoreline was located in the Tyumen and Tomsk regions and partly encompassed the northern parts of the Omsk and Novosibirsk regions. It is not surprising that the Upper Jurassic sand bed $J_1$ in this zone is productive. To date, the greatest number of commercial oil pools in this bed were discovered in the Lar-Yegan and Novy Vasyugan areas of the Tomsk region. However, $J_1$ has not been sufficiently explored, especially along the eastern margin of the basin, where structural pools may also be accompanied by lithologic and stratigraphic pools.

As the result of regression of the Early Cretaceous sea the shoreline shifted somewhat to the northwest, i. e., toward the Shirotnaya Ob region, and this resulted in the creation of a new zone of oil and gas accumulation. The formation of the largest oil deposits in this region can be attributed to the prolonged existence of several paleoriver mouths.

A further marine regression at the beginning of the Cenomanian led to the formation of a new belt of oil and gas accumulation in areas further north; most of the largest gas fields of West Siberia occur in this belt.

By the beginning of 1971 about 70 productive fields had been discovered on the territory being studied. Most of these are oil deposits, the remainder being gas, condensate, and mixed oil and gas. Quite a few oil and gas shows were noted in many horizons in several other regions.

*Samotlor Oil Field*

A major oil field in West Siberia — Samotlor — is situated in the Vakh-Vatinsky Yegan interfluve. It consists of several closely spaced local up-lifts having a common oil-water interface and is confined to the northeastern part of the Nizhnevartovsky arch.

The productive beds of the field, consisting of sheetlike lenticular sand-silt bodies, are separated by clay deposits. The main oil-bearing formations are more than 600 m thick and are of Valanginian-Hauterivian-Barremian and Early Aptian age. The oil pools, mainly of the structural-lithologic type, occur in three groups of strata: $AB_{1-5}$, $BV_8$, and $BV_{10}$.

The group of reservoir beds $AB_{1-5}$ (Lower Aptian-Barremian) occurs below the mudstones of the Aptian stage. These beds have a rather non-uniform structure. Each of them actually consists of alternating sand-stones, siltstones and mudstones which vary in thickness and distribution pattern.

For instance, in well 10 the petroliferous sand layer $AB_{4-5}$ actually con-sists of two parts, 15 and 21 m thick, separated by an argillaceous inter-layer. In well $10^a$, drilled 50 m away from well 10, the sand layer is re-placed by mudstone and is half as thick. The same bed is 47.6 m thick in well 94, while in well 2, 2.5 km to the northwest, it is only 9.2 m thick, i. e., one-fifth as thick.

The distribution diagram of these beds in the Samotlor field (Figure 65) shows changes in the configuration of the sand beds of group AB in time, as well as their interrelationship with areas of argillaceous rocks. Figure 65 clearly reveals that the sandstones show a "meandering" or "patchy" distri-bution pattern. Their formation was undoubtedly controlled by the physio-graphy of the environment of deposition.

The sandstones of $AB_{4-5}$ are polymictic, less frequently arkosic or quartzose. They are relatively well sorted, the grains being subangular or angular, i. e., poorly rounded. The predominant grain size is in the range from 0.1 to 0.25 mm. The cement is chlorite or chlorite-hydromica, less commonly kaolinite. Also present are accessory minerals such as garnet, sphene and epidote, and less commonly anatase. Leucoxene occurs as an authigenic mineral.

Bed $BV_8$ has a fairly irregular distribution and varies in thickness from 20 to 30 m. It consists of fine-grained sandstones belonging to the upper part of the Valanginian stage. Lower down in the section one encounters productive member $BV_{10}$ in which three separate sand beds ($BV_{10}^0$, $BV_{10}^1$ and $BV_{10}^2$) are identified. Each of them is characterized by a different distribu-tion pattern, thickness and reservoir properties.

The entire productive series of Samotlor is characterized by a higher sand content in the eastern and southeastern parts of the field, whereas toward the west and northwest the sandstones are gradually replaced by argillaceous rocks. In the Nizhnevartovsky arch the reservoirs in the east generally have a higher porosity and permeability. All these features are indicative of a supply of detrital material from the east and southeast, as well as of the proximity of the coastal zone.

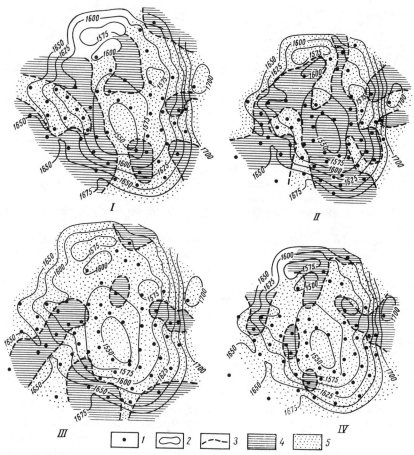

FIGURE 65. Distribution diagram of the productive beds of group AB in the Samotlor oil field (Onishchuk and Sokolovskii, 1971):

Beds: I — $AB_2$, II — $AB_3$, III — $AB_4$, IV — $AB_5$; 1 — wells; 2 — contour lines along the top of a member; 3 — boundaries between lithologic rock types; 4 — mainly argillaceous deposits; 5 — arenaceous deposits.

If one takes into account that in the southeastern part of the West Siberian plate is located the Chulym-Yenisei basin (or the Angara-Chulym depression, according to P. P. Timofeev), in which the paleo-Angara, paleo-Tunguska and paleo-Chulym incised their valleys during the marine regression, it may be assumed that the Samotlor productive series is part of the prodelta deposits of these paleorivers. This field is, therefore, part of one of the largest zones of oil and gas accumulation in the Neocomian belt. There is every reason to believe that other major zones of oil and gas accumulation will be discovered within this belt in localities where large rivers flowed into ancient seas.

*Urenga Gas Field*

The productive series of the unique Urenga gas field is made up of alternating littoral-terrestrial rocks. Gas is present in the upper portion of the Cenomanian, which consists of sand-silt and clay deposits. The sand-silt rocks together account for 50—80% of the section. The thickness of individual sand beds or lenses ranges from 0.4 to 10 m (usually from 2 to 3 m).

The sands and sandstones are composed mainly of quartz (60—68%), feldspar (22—36%), mica (0.2—3%) and rock fragments (2.8—7%). Epidote, sphene and zircon occur as accessory minerals. Abundant authigenic siderite is present as rounded nodules 0.15—1.2 mm in size. The sandy varieties are horizontal or cross-bedded, with intercalations of carbonaceous material or carbonized plant detritus with rare amber inclusions. Interlayers of conglomerates made up of small pebbles have been encountered among the sandstones. The arenaceous rocks often form lenticular bodies. The clay deposits generally consist of a swelling mineral with properties very similar to those of montmorillonite. Kaolinite is of minor significance. There are admixtures of mixed-layer hydromica-chlorite.

Within the section through the Urenga arenaceous-argillaceous deposits local geologists have identified a possible 100-m-thick productive formation which is divided into two parts. The lower part is represented by lacustrine-palustrine, alluvial, deltaic and, less frequently, nearshore-marine facies. The sand-silt varieties contain numerous intercalations of carbonaceous-argillaceous material. In places there are inclusions of carbonized plant remains, fragments of coal and amber.

The upper part of this formation is also characterized by a polyfacies composition, with roughly equal representation of nearshore-marine and continental sediments. V. A. Yezhova has reported an analysis of the cation adsorption complex of the clays in this portion of the section, which indicates that the paleohydrochemical coefficient $(Mg/(Ca + Mg))$ varies from 32 to 54. This coefficient usually has a value of 40—50 under the marine conditions, lower values being characteristic of bodies of fresh water.

Consequently, the hydrochemical sedimentary regime in the region of the Urenga field changed from freshwater to shallow marine.

A study of the grain-size composition of the upper portion of the Urenga productive formation suggests that it was formed in a zone transitional from dry land to shallow sea (Figure 66). This is also suggested by the mineral composition of the clay deposits which contain large amounts of chlorite produced as a result of the abundant supply of iron-containing solutions by river waters. Greater quantities of chlorite are present in the synchrogenic rocks of the Zapolyarnoye and Russkoye fields situated to the east of the Urenga field. Here accumulations of chlorite oolites and leptochlorites occur as lenses and thin intercalations in the conglomerate-like rocks. Glauconite is also present in places.

The nearshore character of the Cenomanian productive formation is confirmed by the fairly common lenticular shape of the sand-silt bodies and by the increase in the sand content in the sections of the Urenga, Medvezhye, Komsomolsky, Vynga-Purovsky, Taz and other fields. This increase can be attributed to the hydrodynamics of wave action during the processes of

sedimentation.   Increased sandiness is apparently characteristic of the
entire coastline of the Cenomanian sea.   The interlayers of small-pebble
conglomerate occurring locally are indicative of river or beach abrasion.

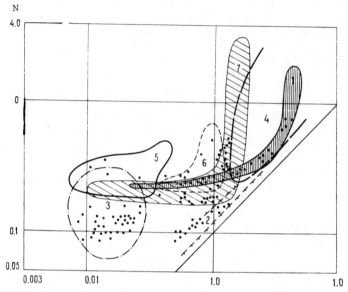

FIGURE 66.   Diagram showing grain-size composition of the sand-silt productive formation
of the Urenga gas field:

Facies:  1 — rivers and permanent ocean currents;  2 — turbidity currents;  3 — calm water;
4 — beach;  5 — calm lagoon;  6 — permanent currents;  7 — marginal parts of shelf.

Variations in the properties of the reservoir rocks due to changes in the
depositional environment can readily be traced in the Zapolyarnoye gas field.
Sand-silt rocks in the eastern part of the structure have a permeability of
1145—2280 millidarcies, whereas in the western part, more distant from the
shore, the permeability drops to 300—460 millidarcies.

The overall thickness of the Cenomanian section in the Urenga area is
300—310 m and increases to 400 m further eastward in the Taz field.   The
sand content increases in the same direction.   Could this not be due to the
presence of the mouth of a paleoriver which debouched into the Cenomanian
marine basin in the Taz region?

The greater part of the Cenomanian deposits (the lower sand-silt member)
in the present-day Nadym-Pur interfluve was formed under conditions of a
coastal lowland plain — in lacustrine-palustrine water bodies, in river valleys, as well as in the alluvial-deltaic zone.   Sarkisyan and Komardinkina
(1971) distinguished, somewhat arbitrarily, two large regions: central Nadym-
Pur (deltaic) and eastern Taz (lacustrine-alluvial).

The first of these regions, which bordered on the marine basin in the west,
was a vast coastal zone covered by river-borne silt-sand deposits of the

prodelta type (Urenga and Nadym-Medvezhye areas). The thickness of sand beds varies from 10 to 15—20 m and they account for 40—72% of the section. Locally, at the base of fine-grained cross-bedded sandstones there occur numerous inclusions of fine (0.3—0.4 cm) poorly rounded dark-gray clay pebbles.

The channel sands of the deltaic sediments cut through synchrogenic silt-clay deposits.

In the Northern arch (the Gubkin, Komsomolsky and Aivasedovo-Pur areas) fluviatile deposits accumulated mainly in small channels of the intradelta.

The mentioned authors believe that the disposition of the ancient hydrographic network was intimately related to the trend of the tectonic structures, which in most cases was meridional. Figure 67 shows the distribution of channel sands in the Gubkin gas field. Channel sands were also encountered in the Taz field (Figure 68) situated to the east of the Gubkin field.

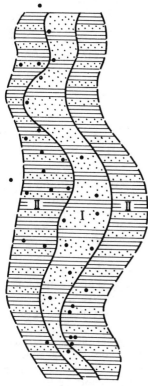

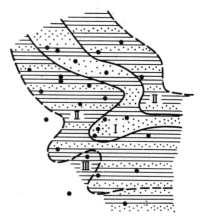

FIGURE 67. Map showing differentiation of a Cenomanian sand-silt member in the Gubkin field (Komardinkina, 1971):

I — arenaceous rocks (50—90%) with a subordinate amount of clay-silt material; II — clay-silt rocks with a subordinate amount of sand material (10—50%).

FIGURE 68. Map showing differentiation of a Cenomanian sand-silt member in the Taz field (Komardinkina, 1971):

I — arenaceous rocks (50—90%) with a subordinate amount of clay-silt material; II — clay-silt rocks with a subordinate amount of sand material (10—50%); III — clay-silt rocks (90—100%).

The thickness of the Cenomanian silt-sand deposits, their coal content and the size of the rock fragments increase in the Taz area due to its proximity to the source area, i.e., the Yenisei Range.  As one approaches the shore-line and the marginal marine depositional plain, the degree of sorting of the sandstones increases and their grain size decreases, the size of the plant detritus decreases, and cross-bedding gives way to horizontal bedding.

The Cenomanian section in the eastern part of the Zapolyarnoye gas field consists of arenaceous rocks (90% coarse silt and sand material).  The content of arenaceous rocks drops to 50—70% further to the west.  Features of these deposits indicate that they accumulated under fairly complex hydro-dynamic conditions, possibly corresponding to a prodelta or nearshore-marine environment.  The physiographic features of the sedimentary basin and the hydrographic network determined the zonation, and in some cases also the channel pattern of the sand and clay-silt sediments.  The distribution of the mouths of major Cenomanian rivers was largely responsible for the formation of the unique, giant gas fields in the northern part of West Siberia.

The configuration of the Cenomanian sea was similar to that of the Early and Middle Jurassic basin, the coastal zone of which could be an independent belt of oil and gas accumulation.  The latter could have served as an additional source of gas for the Cenomanian deposits, but it is also not improbable that it contains its own oil and gas pools.

## 16.  MAIN FEATURES OF PALEOGEOGRAPHIC ZONALITY IN THE DISTRIBUTION OF OIL AND GAS FIELDS IN THE WEST SIBERIAN OIL AND GAS BASIN

The West Siberian Lowlands were an inland sedimentary basin during the Jurassic and Cretaceous.  Numerous rivers supplied it with detrital and chemogenic material.  However, the lithofacies and paleogeographic maps prepared by individual investigators and teams of scientists (V. P. Kazarinov, N. N. Rostovtsev, T. F. Balabanova, S. G. Galernin, T. I. Gurov and others) usually only show the general facies distribution and the chief features of the paleogeographic environment of the stratigraphic divisions or, at best, stages.  Lacustrine-palustrine, coastal plain, lagoonal, nearshore-marine and marine facies zones are distinguished.  Sometimes fluviatile and even deltaic facies are shown.  However, the exact location of the delta regions has not been established as yet.

For the Early Jurassic period Sarkisyan, Korzh et al. (1968) were only able to indicate hypothetical river valleys in the alluvial-lacustrine plain that occupied the central part of the West Siberian basin.  Novosibirsk geologists have established the presence of deluvial-proluvial and fluviatile formations in the Lower and Middle Jurassic deposits of the Ob-Irtysh interfluve; the fluviatile deposits are characterized by the linear distribution of the mainly arenaceous rocks which extend as a distinctive band from south-east to northwest.

Since it is possible to locate and sometimes even trace fluviatile deposits, it follows that they must terminate in deltas, estuaries or bays whose sediments, being situated at a lower hypsometric level, are better preserved in

the fossilized state than the fluviatile sediments.  Although none of the
paleogeographic maps depicts even a single deltaic zone, it is to be hoped
that, as additional drilling data and geologic information on the West Siberian
basin become available, it will be possible to distinguish and delineate large
paleodeltas.

During the Early Jurassic most of the area of the West Siberian plate was
a humid plain with a well-developed hydrographic network.  A marine basin
existed only in its northeastern part, a gulf of which penetrated into the alluvial-lacustrine swamped lowland.  During the Middle Jurassic the sea
advanced southward, approximately to the latitude of the present-day lower
reaches of the Irtysh.  The marine ingressions during the Bajocian and partly also the Bathonian encroached deeper into the southwestern regions.

By the end of the Middle Jurassic only vast relict lagoons remained in
these regions, in which variegated sediments accumulated.  The lagoons
were of the freshwater type, indicating a large influx of river waters.

The Late Jurassic is characterized by an extensive marine transgression
which covered a considerable part of the region in question.  The sea was
shallow with individual islands or even archipelagoes.  The northeastern
regions and the Mansi syneclise, the Pur megadepression and other depressions were the deepest parts of this sea.  The size of the inundated territories was in all probability quite variable.  Therefore, the transition from
continental to marine environments covers a rather broad zone.  In order
to establish the position of the mouths of the major Early and Middle Jurassic rivers, it is first necessary to determine the main direction of transport of the terrigenous material into the sedimentary basin.  Several investigators of West Siberia have attempted to tackle this problem.  V. P. Kazarinov, E. V. Shumilova, T. I. Gurova, M. Ya. Rudkevich, B. V. Topychkanov,
T. A. Yastrebova, S. G. Sarkisyan and others have identified terrigenous-
mineralogic provinces of individual suites and pinpointed the source areas.
Most geologists regard the Urals as the provenance for the western part of
the basin.  However, recent studies have established that a meridional rather
than a latitudinal direction of supply of terrigenous material is particularly
evident in the Lyapin trench.  The trend of decrease in the content of unstable minerals and increase in the amount of stable minerals, indicative of
the direction of the main streams which transported the detrital material,
is from south to north both in the Lyapin trench and in the Mansi syneclise.
This suggests that the Urals were not the principal provenance and that they
were a relatively low-lying territory at that time.

The range and content of unstable heavy minerals gradually decrease
from east to west.  Minerals such as apatite, staurolite, kyanite and sphene
are encountered in the eastern and, especially, southeastern regions; they
are not found in the west.

The detrital material derived from the Urals did not spread beyond the
Lyapin trench or other depressions adjacent to the Urals.  The meridional
direction of transport of the terrigenous material in the Lyapin trench is
predominant, and the content of unstable minerals decreases gradually from
south to north, from 42—36% in the Ust-Manya region to 12.5% in the Saran-
paul region.  A similar phenomenon is noted in the Mansi syneclise which
was supplied with detrital material from the east and then from the north.
At the same time, considerable amounts of detrital material were supplied

from the south.   For instance, south of Tyumen, in the Ryaziev area, 64% of the heavy minerals are stable in the Early and Middle Jurassic deposits, while to the north of Tyumen the figure increases to 68%, in the Cherkazinsky area to 80%, and in the Dobrinsky area to 90%.

These data show that the main channels of transport of the terrigenous material ran approximately along the upper and middle reaches of the Ob, as well as along the Irtysh.   The present-day valleys of these rivers are, in all likelihood, inherited from the ancient valleys of the major Jurassic and Cretaceous waterways.

Widespread among the rocks of the Lower and Middle Jurassic, as well as of the Neocomian, are sand-silt formations which serve as reservoirs of oil and gas.   In terms of origin they are classified as fluviatile and near-shore-marine deposits and, less frequently, as sediments of relatively calm water bodies.   The genetic features of these deposits are cross, wave-cross, criss-cross, interrupted and horizontal bedding.   The arenaceous members frequently contain mudstone intercalations of floodplain or lagoonal facies. Locally, one also encounters what appear to be beach sandstones showing indistinct bedding.

The grain size of the arenaceous rocks is also quite variable, ranging from gravelites to fine-grained varieties with the latter predominating. Medium-grained sandstones are less common.

Inhomogeneity is the most characteristic feature of the arenaceous rocks penetrated by boreholes in the West Siberian basin.   The coarsest clastic material is encountered in basal horizons marking the onset of transgressive cycles of sedimentation.   These sediments were usually deposited on alluvial coastal plains and in littoral zones of the shallow seas not only fringing the continent but also surrounding certain large and small islands.   Many oil and gas pools occur in these horizons.

We have already mentioned that deltaic deposits of paleorivers originating in the Angara-Chulym depression were very common in the southeastern part of the West Siberian Lowlands during Jurassic and Cretaceous times. Timofeev (1970) divides this depression into three sectors from a structural-morphologic point of view: southeastern, northeastern, and southwestern.   The paleo-Angara, the largest river, passed through the first sector.   The northeastern sector was drained by the paleo-Tunguska, while the southwestern sector was traversed either by the paleo-Chulym itself or by its numerous tributaries, the confluence of which produced a rather large river.   All these ancient rivers debouched into a marine basin of variable area, the shoreline of which alternately approached and withdrew from the Angara-Chulym depression.   Each significant change was naturally followed by changes in the sedimentary environment and in the position of the belts and major zones of oil and gas accumulation.

The delineation of estuarine zones of paleorivers and the tracing of the deposits of river channels, bars and other coastal sand formations must become a principal guideline in oil and gas prospecting in the southeastern part of the West Siberian oil and gas basin.   The rivers usually formed vast alluvial-deltaic plains in their lower reaches and branching subaqueous prodeltas in the sea.   Repeated shifts of the shoreline also had a considerable effect on the formation of productive traps.

All these features governing the distribution of oil and gas pools can be
revealed only by way of a thorough paleogeographic analysis.

Fluviatile sandstones possess good reservoir qualities.  However, they
are quite variable.  In the Early and Middle Jurassic section of the Ust-
Silgin, Sredny Vasyugan, Novy Vasyugan and other regions, one notes that
fluviatile deposits are followed by floodplain or palustrine-lacustrine deposits and then reappear once again higher up.  In some areas of the West
Siberian basin oil and gas are associated with the basal sand horizon of the
Upper Jurassic transgression.

A mere glance at the map showing the oil and gas fields discovered in
this enormous and complex basin readily reveals that the distribution of
most of them is according to a definite pattern.  A group of large oil fields
extends along the Middle or Shirotnaya Ob region.  Another chain of smaller
fields is located along the western margin of the Mansi syneclise.  Finally,
the largest gas fields occur in the northern regions.

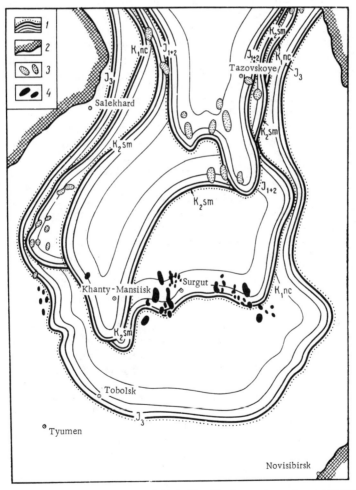

FIGURE 69.  Distribution diagram of oil and gas fields in West Siberia superimposed on a paleogeographic map:

1 — outlines of ancient coastal zones;  2 — boundary of the Paleozoic;  3 — gas fields;  4 — oil fields.

When this map showing the distribution of the productive fields is super-imposed on a paleogeographic map featuring the postulated positions of shore zones of the Early and Middle Jurassic, Late Jurassic, Neocomian and Ceno-manian marine basins (Figure 69), one notes that the vast majority of the fields tend to lie close to these zones, i. e., they are located in belts of oil and gas accumulation. The largest fields in the northern regions of West Siberia occur in the Cenomanian and Early and Middle Jurassic coastal zones, although no gas has been discovered in the Cenomanian to date. It is quite possible that belts and major zones of oil and gas accumulation will yet be discovered within the Lower and Middle Jurassic nearshore-marine sediments in northern parts of the Tyumen region. The latter may contain oil rather than gas, supportive evidence being the fact that nearly all the major gas fields are associated with newly formed structures, whereas the oil fields occur in inherited structures.

Consequently, belts of oil and gas accumulation associated with paleogeo-graphic zones are readily distinguished within the West Siberian megabasin. It is still difficult at present to pinpoint major zones of hydrocarbon accumu-lation due to the lack of sufficient geologic and other data. However, even the available material makes it possible to delineate them tentatively. The Shirotnaya Ob region, in which were located the mouths of large Late Jurassic and Neocomian rivers, falls in this category. Such rivers possibly include the paleo-Angara, paleo-Ob, paleo-Irtysh and other ancient waterways.

The Ob-Irtysh interfluve, which included the Shirotnaya Ob region at that time, is part of the vast transition region from the alluvial-deltaic plain to the seacoast with its widely developed prodelta formations. For instance, the Neocomian section of the southern and southeastern parts of the inter-fluve consists mainly of alluvial and alluvial-deltaic facies. Further north, closer to the right bank of the Ob, nearshore-marine facies prevail. They are characterized by frequent variations and transitions from sand-silt varieties to clay varieties. Shallow-marine facies characterized by a rela-tively persistent distribution of arenaceous rocks alternating with argilla-ceous rocks occur along the right bank of the Shirotnaya Ob.

The intensity of sediment accumulation in the Shirotnaya Ob region was probably promoted by the processes occurring near the mouths of the rivers. These resulted in the accumulation of sand formations extending in a sub-meridional direction along the coastal zone. Major oil fields occurring in prodelta deposits are associated with these formations (the Samotlor, Ust-Balyk, etc.).

The giant Samotlor oil field discovered on the Nizhnevartovsky arch is con-fined to the Hauterivian-Barremian and Lower Aptian deposits which are very similar in character to typical prodelta formations. The explored part of the productive section alone contains up to 15 petroliferous beds with high porosity and permeability.

It seems probable that more than one paleomouth was located in the Shi-rotnaya Ob region. There is every reason to believe that detailed recon-struction of the paleogeographic environment in which this major zone was formed will make possible the discovery of other large oil fields. Not only structural traps hold promise here; lithologic and stratigraphic traps should also be taken into account.

Belts of oil and gas accumulation in West Siberia extend as a broad band along the shores of marine basins which existed for varying time intervals. The most stable coastal zones should become the major objects in the search for zones of oil and gas accumulation.

The southern part of the Surgut arch is undoubtedly one such zone. Several large oil fields have already been discovered there (Ust-Balyk, Bystrin-Vyngon-Minchimkin, Fedorovskoye, etc.). The total number of sand beds in the Early Cretaceous section here varies from 10—20 in the northwestern regions to 35—45 in the southeastern regions. A major zone of oil accumulation is associated with the Nizhnevartovsky and Aleksandrovsky arches, a considerable part of which was located in the estuarine zone of a large paleoriver in Late Jurassic and Early Cretaceous times. Major zones of oil and gas accumulation may also occur in other regions of West Siberia, in particular in the northwest where the broad alluvial valley of a large Early Jurassic paleoriver has been identified. The prospects for discovering major zones along the east coasts of the Jurassic and Cretaceous marine basins are also promising.

In northern regions the increase in thickness of the Zavodoukovsk series is accompanied by an alternation of the marine and nearshore-marine facies. However, the Jurassic section here has been explored only by isolated boreholes and is still quite poorly studied. Nevertheless it is of great practical interest, since estuarine zones of large paleorivers with rich oil and gas accumulations may exist in the coastal belt of the Early and Middle Jurassic marine basin.

General geological assumptions, together with the available data on the better-studied southern regions, make it possible to suggest the most probable disposition of the banks and paleodeltas of the major Early and Middle Jurassic rivers (Figure 70). The upper reaches of one such river were at one time located within the Turgai plateau which was then a plain covered by lakes and marshes. Further north, the river valley, which was fringed by elevated forms of the ancient relief in the west and east, passed through the Chelyabinsk graben where coal accumulation was controlled by paleogeographic rather than tectonic factors.

The central part of the graben was filled with fluviatile, floodplain, and swamp-oxbow formations, while the valley itself, which followed a submeridional course, became broader toward the north. In the west it was probably fed by smaller tributaries which carried fine-grained alluvium from the Urals. Since coal-free sediments of alluvial cones occur in the eastern part of the main river valley, the right banks of the paleorivers were evidently higher than the left banks.

Thick lenses of clarain, clarodurain, and duroclarain accumulated in the numerous, rather deep oxbows and lakes located in the broad river valley. These coal varieties sometimes contain lenticular inclusions of vitrinite or fusinite which are responsible for the banded or striated structure. This composition suggests that the coal was formed from finely comminuted material deposited in stagnant or nearly stagnant waters. Wood, bark, and leaves were generally the starting material for the clarain coals. Vitrinite and fusinite were formed mainly from large wood fragments that had undergone various transformations. The sand-clay material encountered in some fragments of tree trunks suggests that these fragments were introduced by streamflow.  The presence of discontinuous intercalations and with pinch-outs and of clay and sand gouges among the vitrain varieties is also indic-ative of a regime of running water.  Moreover, the splits and pinch-out of the coal seams point to a change in the environment of coal accumulation, which is usually related to changes in the hydrodynamic regime of river waters flowing near or debouching into lakes.

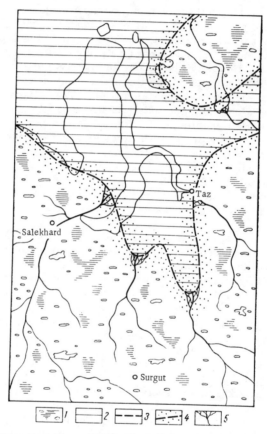

FIGURE 70. Paleogeographic map of West Siberia during the Early and Middle Jurassic:

1 — lowland covered with marshes and lakes; 2 — marine basin; 3 — shoreline; 4 — zone most favorable for oil and gas accumulation (belt); 5 — assumed location of large rivers and their deltas.

Nearly all coal-bearing deposits on the eastern slope of the Urals overlie rudaceous formations of alluvium, ephemeral streams, alluvial bars, etc., i. e., features which are typical of erosion surfaces of rocks of different age. These features are characteristic of limnetic deposits which accumulated in areas remote from the seacoast.

The coal-bearing deposits of the Urals' eastern slope suggest the prolonged existence of continental conditions of sedimentation along the western and southwestern margins of the West Siberian Lowlands.  During the Early and Middle Jurassic the relatively low Ural Mountains became a hilly region with a fairly dissected relief resulting both from tectonic movements and erosion processes.  Deluvial and proluvial sediments were deposited on the gentle slopes and at the bottom of the hills, while alluvium accumulated in the valleys separating them.  The low-lying forms of the ancient relief were occupied by numerous lakes and marshes in which the organic source material for coal formation accumulated.  Further away from the Urals, in the direction of the central part of the West Siberian Lowlands, was an alluvial plain covered by both small and fairly large freshwater bodies and river valleys.  A marine basin existed only to the northeast and north of the lowland.

The paleogeographic zonality had a pronounced effect on the distribution pattern of the oil and gas reservoirs in the West Siberian Lowlands.  In contrast to the East European Platform, where oil- and gas-bearing regions stretch along the downwarped margins, the marginal parts of the West Siberian Lowlands are virtually devoid of oil and gas accumulations.  The latter are widespread at a certain distance from the fringes of the lowland and increase in frequency toward its center which underwent considerable subsidence.

Oil and gas fields in the western part of the lowland are, as a rule, associated with basal arenaceous rocks which usually pinch out toward elevated forms of the ancient relief.  The regional pinch-out zone extends along the entire eastern slope of the Northern Sosva Range.  In this connection, the distribution of the arenaceous formations, which was determined by the paleogeographic environment and, primarily, by the position of the coastal zone and the hydrodynamics of the river system, is of great practical interest.

The oil and gas pools in the Berezovo, Shaimsky, Krasnoleninskaya and other western regions are controlled by the pinch-out of sand deposits, and, therefore, lithologic traps must predominate here.  The tops of local structures are often found to lack sandstones, and the latter occur as elongated lenticular or channel bodies on the flanks of uplifts or beyond them.  In such cases one may speak of lithologic pools whose outlines usually do not coincide with the closed part of the structure.

These monoclinal sandstones may be productive beyond the limits of local positive structures.  The extent of their productivity is probably dependent on their thickness and their contacts with source beds.  Areas where thick sandstones alternate with argillaceous members are the most favorable sites for oil and gas accumulation in both structural and nonstructural traps.  In most cases these are areas of paleodelta deposits.  Other sand deposits of alluvial (fluviatile) or nearshore-marine formations (spits, bars or beaches) can also serve as good reservoirs.

Karagodin (1966) points to the fluviatile character of the reservoirs in the Krasnoleninskaya arch region.  The continental deposits of the Tyumen series in this region attain a thickness of more than 100 m and pinch out only in the upper parts of individual uplifted areas.  The horizon containing commercial amounts of oil shows a fluviatile facies.  For instance, well 13, located between the northern and central elevations ("bald patches") of the Kamenny uplift, penetrated a productive horizon more than 20 m thick. It is made up of inequigranular sandstones alternating with gravelites and conglomerate interlayers. The abundant carbonaceous detritus accentuates the cross-bedding which is predominant in the sandstones. The clastic material is rather angular, usually poorly sorted. In well 12 in the same area the base of the section consists of gravelite and conglomerate beds in which the rock fragments are no larger than 2—4 cm.

The sand formations discovered in the Krasnoleninskaya arch occur in canyonlike valleys of the ancient relief and are thought to be of proluvial and fluviatile facies. Since the deposits of these paleorivers are confined to tectonic and erosion-produced depressions, they should be sought beyond the limits of the upper portions of present-day uplifts which are inherited to a certain extent from older structures. We are in full agreement with Yu. N. Karagodin that the oil pools in the Krasnoleninskaya region will prove to be of a channel-shoestring type and that the oil-water contact will be at some distance from local uplifts.

If the increase in thickness and improvement of reservoir properties of the producing formations in the western part of the West Siberian Lowlands are connected with the action of paleorivers which incised their courses through tectonically weakened or geomorphologically depressed zones, then oil and gas accumulations may occur not only in structural traps, but also in lithologic and stratigraphic traps. The pools will occur mainly in lenticular or channel sands.

The thickness of the Late Jurassic productive sand-silt rocks varies from 40—50 to 600—700 m, accounting for 5—10 to 60% of the entire series. The greatest thickness is encountered in the northeastern part of the lowland, especially in the Ust-Yenisei depression. The paleogeographic environments of deposition of these series are, unfortunately, poorly studied to date.

Arenaceous rock types are fairly widely developed in the Early Valanginian formation, with the exception of the western regions. The sand deposits of the eastern and central regions are discontinuous. They usually occur as lenticular bodies several kilometers in length and up to 10 m, less commonly 20—25 m, thick. Sandstones account for 5—10 to 50—60% of the section; they are of minor significance west of the Khanty-Mansi meridian but their content increases in areas adjacent to the Taimyr Peninsula.

The following productive formations, which include the Late Valanginian and Hauterivian-Barremian, and partly also the Aptian, oil- and gas-bearing complexes, contain the major oil fields of the Shirotnaya Ob region. At that time the coastal zone of a somewhat less saline shallow-water marine basin existed here. Sand-silt rocks become more abundant and their reservoir qualities improve toward the east and southeast; the sandstones are usually channel-shaped or lenticular. The mineral composition of the rocks indicates that the detrital material was brought mainly from the southeast.

The Hauterivian-Barremian was characterized by a revival of tectonic movements, a certain regression of the sea and the more rapid supply of the terrigenous material. Variegated sediments of lagoonal, lacustrine and fluviatile facies were deposited in the southern and southeastern parts of the basin. Gray sediments of marine origin accumulated in the central and northern parts. Figure 71 is an attempt to depict the coastal zone of the Early Cretaceous sea and the lower reaches of the paleorivers which debouched into it. The map is only tentative and requires further refinement.

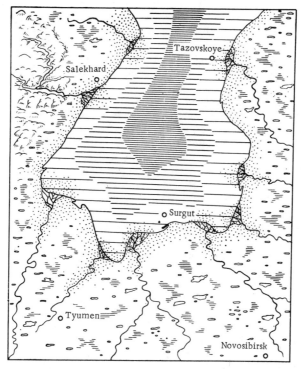

FIGURE 71. Paleogeographic diagram of West Siberia during the Early Cretaceous

The considerable decrease in salinity and the character of the sand reservoirs in the oil fields already explored indicate that several large rivers debouched into the southern part of the sea. The largest oil pools of the Shirotnaya Ob region were generally discovered in estuarine, mainly prodelta, multilayered lenticular or channel sands.

The Aptian-Albian-Cenomanian productive formations are composed of continental and nearshore-marine facies which locally show transition to variegated, lagoonal and deltaic facies. The enormous gas fields of West Siberia occur in arenaceous rocks of these formations; several oil pools have also been discovered here. We shall not go into details of the mineralogy, petrography and grain size of the rocks of the Cenomanian gas fields in the north of Tyumen, and will only discusss the sedimentary environment of one of them — the Gubkin gas field situated in the southeast of the enormous, largely gas-bearing territory which is part of the Yamal-Gyda syneclise. The most characteristic feature of the arenaceous rocks on the Gubkin (Purpe) uplift is that they are well sorted and of coarser grain size in its anticlinal part, whereas on the slopes there is an increase in the quantity of fine-grained siltstones giving way to clays. The sand and clay deposits have a pronounced zonal distribution. The meridional band of predominantly sand-silt rocks is replaced by argillaceous deposits toward the west.

The paleogeographic environment of the Yamal-Gyda syneclise during the Cenomanian is pictured as follows: the central part of the syneclise was occupied by a semiclosed marine basin of low salinity, in which relatively shallow-water sediments rich in organic matter were deposited. A broad lowland covered by swamps, lakes and alluvial deposits fringed the basin to the west, east and southeast. The character of the facies within the Pur-Taz interfluve region indicates that the lowland was repeatedly inundated by the sea during the Cenomanian.

Detrital material was carried into the marine basin by numerous rivers. In the Gubkin area mainly fine-grained sands accumulated. Widespread among them are siltstones with an admixture of argillaceous rocks and showing characteristic unidirectional, slightly oblique or wavy-horizontal bedding. Frequent changes in the lithology and textural features of the deposits along the strike and in section, as well as differences in the sorting of the material, are indicative of the variable hydrodynamic force and direction of the currents, which are typical of coastal zones of marine basins.

A study of the grain size of the rocks making up the Cenomanian productive horizons suggests that sedimentation in the Gubkin area took place on a shallow-water shelf under conditions corresponding to those of a developing lagoon, with variable offshore currents and surf.

The presence of considerable amounts of neogenic siderite in the rocks indicates that the salinity of the water in the basin had decreased, apparently owing to the influx of river waters. This is also suggested by the high content of kaolinite. The deposition of the lower, arenaceous member of the Cenomanian deposits in the Gubkin area took place both in the coastal part of the sea and in semiclosed water bodies of the lake-lagoon type, as well as in the estuarine parts of quietly flowing lowland rivers. A deep-water sedimentary environment is typical of the upper, clay-silt member. The terrigenous material supplied to the territory under study was derived from the east and northeast. This is confirmed, in particular, by the increase in the sand content and the quantity of polymictic arkosic varieties in the direction opposite to that of transport.

Thus, in this case too, we are dealing with a favorable nearshore-marine paleogeographic environment of deposition of the gas-bearing formation. In connection with the Cenomanian productive complex, it should be pointed out that, following the regional regression which began at the end of the Early Cretaceous, the outlines of the Cenomanian marine basin were very similar to those of the Early and Middle Jurassic sea. It is thus reasonable to suppose that the Early Jurassic belt of oil and gas accumulation essentially coincides with the Cenomanian belt.

There is no doubt that new major oil and gas fields will be discovered within the enormous West Siberian oil and gas basin. The experience gained by the Soviet and non-Soviet scientists shows that the most favorable conditions for the formation of such fields existed in estuarine zones of large paleorivers which flowed in plains. It is certainly not easy to identify formations of ancient deltas which characteristically feature changes in sedimentary environments over relatively small areas. Changes in the tectonic regime and climate caused rearrangement of areas with different facies environments, leading to a patchwork distribution pattern of the deltaic deposits.  Recognition of the latter is a complicated task because most parts of West Siberia have been insufficiently covered by boreholes, even in regions where oil and gas fields have already been discovered.

Further drilling operations and research work will provide a basis for a more detailed reconstruction of the physiographic environments in which the productive formations were formed, and for delineating paleoriver mouths and establishing the location of belts and major zones of oil and gas accumulation.

It thus follows that detailed paleogeographic studies are a most important prerequisite for the discovery of new large oil and gas fields in West Siberia.

## CONCLUSION

Summarizing the results of studies and generalization of the factual data on the distribution of commercial accumulations of hydrocarbons both in the Soviet Union and in other countries, we may conclude that paleogeography — the science of former physical and biogeographic environments during the evolution of the geosphere — is no less important as a basis for oil and gas exploration than tectonics, geochemistry and other earth sciences. Paleogeography is an integral part of the system of scientific disciplines constituting the theoretical basis for exploration for hydrocarbons.

The numerous facts discussed in this book point to the importance and timeliness of paleogeographic studies for petroleum and gas geology. The methods and procedures of these studies mainly involve comprehensive investigation of core samples and various well logs. These methods can be used by a wide circle of scientists and prospectors, being both simple and cheap. Their application in practice will undoubtedly lead to new discoveries and will make it possible to save millions of rubles wasted on "dry" wells. The paleogeographic method, together with the achievements of lithogenesis (historical, geologic and geochemical methods of assessing the oil and gas content in sedimentary basins) and other branches of geologic science, broadens the scientific basis and practical possibilities of oil and gas exploration.

The revealed association of commercial accumulations of oil and gas with coastal zones (belts) and of large fields with estuarine regions of paleorivers (major zones of oil and gas accumulation) is an expression of the laws governing the distribution of these fossil fuels within the earth's crust. Practical utilization of this established regularity will make it possible to:

1) forecast and conduct special exploration for large oil and gas fields, proceeding from the paleogeographic sedimentary environments of the producing formations;

2) reevaluate the oil and gas prospects of particular territories and select the most effective direction of exploration;

3) make a better substantiated choice of structures for deep exploration based on their position relative to belts and major zones of oil and gas accumulation;

4) direct to best advantage exploration for nonstructural types of pools occurring in sedimentary formations of channels and distributaries of paleorivers, bars, spits, and other sand bodies of coastal and estuarine zones;

5) present new tasks to be tackled by geophysical methods of exploration in order to identify and trace the lithofacies complexes of ancient deltas, bars, reefs and other formations;

6) find optimal systems for exploiting productive fields, proceeding from genetic characteristics of the reservoir rocks established by detailed paleogeographic reconstruction.

This is the gist of the paleogeographic principles of oil and gas prospecting.

In this book special emphasis is laid on the fact that many unique and large fields occur in formations of ancient deltas and prodeltas which developed under conditions of steady subsidence. In 1955 for the first time we were able to identify the estuarine regions of the Lower Carboniferous paleorivers in the Transvolga region near Kuibyshev and to reveal the key role of the very favorable combination of factors in the formation of large oil accumulations. Later, several other formations of ancient deltas containing large oil pools were discovered in the eastern and southeastern parts of the East European Platform. As already mentioned, a similar pattern has been noted in West Siberia, as well as in many oil and gas basins of the U. S. A., Canada, Venezuela and other countries.

The rather significant conclusion that the estuarine deposits of ancient rivers are highly productive is not a subjective judgment based on individual studies, but rather an objective reality corroborated by practice. It would be an unmitigated error not to utilize this achievement of scientific thought.

It seems very desirable to organize laboratories, departments, or groups dealing with paleogeographic analysis in the respective research and prospecting establishments. Colleges training petroleum geologists could introduce a special course into their programs: methods of paleogeographic studies in geologic exploration. These measures might contribute to a better understanding of the significance of paleogeography and an awareness of its possibilities, fostering a paleogeographic outlook in a wide circle of geologists.

A further increase in proved reserves and oil and gas production does not seem possible without the introduction of new trends and ideas into petroleum geology, in particular paleogeographic methods. If geologists are either unaware of or disregard this kind of information, they will be unable, in the absence of uplifts, to adopt a correct approach to the identification of lithologic and stratigraphic pools and to achieve optimum results from drilling. Paleogeographic reconstructions must become an integral part of the entire complex of studies determining the ways and means of scientific prediction and practical application of oil and gas exploration.

They call for minimum equipment, but maximum initiative in making the most of the factual data available. This, however, is not all. It is not sufficient merely to reconstruct the physiographic sedimentary environment of producing formations and prepare a series of paleogeographic maps. One must also be able to put them to proper use in solving practical problems.

Also essential for scientific and technological progress in petroleum geology is the speeding up of the application of theoretical principles. We can and must discover oil and gas fields on new and long-worked territories by the application of new concepts and hypotheses, especially those already corroborated by practice.

The prime importance of the paleogeographic trend in the theory and practice of petroleum exploration must be recognized.

# BIBLIOGRAPHY*

ALEKSIN, A.G., I.Kh. ABRIKOSOV, V.A. GROSSGEIM, et al. Lithologic and Stratigraphic Oil and Gas Pools: Methods of Exploration. — In: Stratigraficheskie lovushki. Moscow, Natsional'nyi Komitet SSSR po Nefti, 1971. (Russian)

ANDEL, T.H. Sediments of Rhome Delta, p.169. — New York, 1958.

ARONOW, S. Nueces River Plain of Pleistocene Beaumont Formation, Texas. — Bull. Am. Ass. Petrol. Geol. **55**, No.8 (1971), 1231—1248.

AIZENBERG, D.E. Stratigraphy and Paleogeography of the Lower Carboniferous in the Western Sector of Greater Donbass. — Kiev, Akad.Nauk Ukr SSR, 1954, 272 pp. (Russian)

BAIDIN, S.S., F.N. LINBERG, and I.V. SAMOILOV. Hydrology of the Volga Delta. — Moscow, Gidrometeoizdat, 1956. (Russian)

BAKIROV, A.A. Oil- and Gas-Bearing Regions of North and South America. — Moscow, Gostoptekhizdat, 1959, 296 pp. (Russian)

BAKIROV, A.A., M.I. VARENTSOV, and E.A. BAKIROV. Oil and Gas Provinces and Regions Outside the Soviet Union. — Moscow, "Nedra," 1971, 541 pp. (Russian)

BATES, C.C. Rational Theory of Delta Formation. — Bull. Am. Ass. Petrol. Geol. 37, No.9 (1953), 2119—2162.

BATURIN, V.P. Paleogeography according to Terrigenous Components. — Moscow, Otd. Nauch. Tekhn. Informatsii, 1937, 292 pp. (Russian)

BATURIN, V.P. Petrographic Analysis of the Geologic Past according to Terrigenous Components. — Moscow, Gostoptekhizdat, 1947, 339 pp. (Russian)

BENTHAM, R. Oil in Burma. — Inst. Petrol. Rev., No.236 (1966), 148—160.

BOTVINKINA, L.N. Bedding of Sedimentary Rocks. — Moscow, Akad. Nauk SSSR, 1962, 541 pp. (Trudy Geol. Inst. Akad. Nauk SSSR, No.59). (Russian)

BOTVINKINA, L.N. Methodological Handbook on the Study of Bedding. — Moscow, "Nauka," 1965, 259 pp. (Russian)

BROD, I.O. Modern Views on Formation and Distribution Patterns of Oil and Gas Accumulations. — Geologiya Nefti i Gaza, No.11 (1960), 1—8. (Russian)

BROD, I.O. Main Distribution Patterns of Global Oil and Gas Accumulations. — In: Spisanie na b"lgarskogo geologicheskogo druzhestvo, No.3, pp.1—14. Sofia, 1960. (Russian)

BUSCH, D.A. Prospecting for Stratigraphic Traps. — Bull. Am. Ass. Petrol. Geol. **43**, No.12 (1959), 2829—2843.

BUSCH, D.A. Methods of Prospecting for Stratigraphic Oil and Gas Traps. — Bull. Ass. fr. Techns. Pétrole, No.161 (1965), 167—181.

BUSCH, D.A. Genetic Units in Delta Prospecting. — Bull. Am. Ass. Petrol. Geol. 55, No.8 (1971), 1137—1154.

CARRIGY, M.A. Some Sedimentary Features of the Athabasca Oil Sands. — Sedim. Geol. 1 (1967), 327—352.

CARRIGY, M.A. Deltaic Sedimentation in Athabasca Tar Sands. — Bull. Am. Ass. Petrol. Geol. 55, No.8 (1971), 1155—1169.

CARSEY, J.B. Geology of Gulf Coastal Area and Continental Shelf. — Bull. Am. Ass. Petrol. Geol. **34**, No.3 (1950), 361—385.

CHEPIKOVA, K.R.M. (Editor). Geology and Oil- and Gas-Bearing Reef Complexes. — "Mir." (Russian translation, 1968)

CHOCHIA, N.G. Geologic Structure of the Kolva-Vishera Territory. — Moscow, Gostoptekhizdat, 1955, 407 pp. (Russian)

CHOLOVSKII, I.P. Methods of Geologic and Economic Analysis in the Development of Large Oil Fields. — Moscow, "Nedra," 1966, 180 pp. (Russian)

CONYBEARE, C.E.B. Oil Accumulation in Alluvial Stratigraphic Traps. — Aust. Oil and Gas J., No.11 (1964), 3—8.

CONYBEARE, C.E.B. Origin of Athabasca Oil Sands. — Bull. Can. Petrol. Geol. **14**, No.1 (1966), 84—91.

* [The reader is advised that the non-Russian items in the Bibliography have been reproduced faithfully from the Russian text, aside from the correction of obvious typographical errors.]

CRAFT, W.E.  Channel Sands Are the Key to Wilcox Oil. — Oil Gas J., **64**, No.15 (1966), 111–118.

DAVIS, J.A.  Offshore Areas of the World Continue to Draw Oil Hunters. — Oil Gas Inter., No.12 (1966), 127–135.

DENNISON, J.M.  Petroleum Related to Middle and Upper Devonian Deltaic Facies in Central Appalachians. — Bull. Am. Ass. Petrol. Geol. 55, No.8 (1971), 1179–1193.

DICKEY, P. and R. RON.  The Influence of Facies on Oil Distribution. (Russian translation, 1961)

EGOROV, A.K.  Coal Belts and Oil- and Gas-Bearing Zones of the World. — Rostov. Univ., 1960, 183 pp. (Russian)

EMERY, K.O.  Atlantic Continental Shelf and Slope of the United States, p.266.

EMERY, K.O.  Sediments and Water of Persian Gulf. — Bull. Am. Ass. Petrol. Geol. **40**, No.10 (1956), 2354–2383.

EMERY, K.O.  Relict Sediments on Continental Shelves of the World. — Bull. Am. Ass. Petrol. Geol. **52**, No.3 (1968), 579–594.

FALINI, F.  On the Formation of Coal Deposits of Lacustrine Origin. — Bull. Am. Ass. Petrol. Geol. **49**, No.12 (1965), 1317–1346.

FISK, H.N.  Fine-Grained Alluvial Deposits and Their Effects on Mississippi River Activity. — Miss. River Comm., 1947, 82 pp.

FISK, H.N. and E. McFARLANE.  Late Quaternary Deltaic Deposits of the Mississippi River. — Spec. Pap. Geol. Soc. Am., No.62 (1955), 279–302.

GARDNER, F.J.  Offshore Oil — the Search Now Spans Globe. — Oil Gas J. **64**, No.25 (1966), 109–120.

GARDNER, F.J.  Surprising Nigeria Oil Flow Surpasses All Predictions. — Oil Gas J. **69**, No.14 (1971), 29–32.

GERASIMOV, I.P. (Editor).  Application of Geomorphological Methods in Structural Geology Studies. — Moscow, "Nedra," 1970, 296 pp. (Russian)

GORETSKII, G.I.  Alluvium of Large Anthropozoic Paleorivers of the Russian Plain. — Moscow, "Nauka," 1964, 415 pp. (Russian)

GORETSKII, G.I.  Alluvial Record of the Great Paleo-Dnieper. — Moscow, "Nauka," 1970, 490 pp. (Russian)

GOLOSOV, S.P., Yu.A. DOLGOV, and V.I. MOLCHANOV.  Physicochemical Changes in Minerals on Superfine Comminution. — Novosibirsk, Sibirsk. Otd. Akad. Nauk SSSR, 1966, 135 pp. (Russian)

GOSTINTSEV, K.K. and V.A. GROSSGEIM.  Stratigraphic and Lithologic Oil and Gas Pools. — Leningrad, "Nedra," 1969, 364 pp.; Trudy Vses. Nauch.-Issled. Geologo-Razvedoch. Inst., No.280, 1969. (Russian)

GROSSGEIM, V.A.  Bottom Currents and Tectonics. — Sovetskaya Geologiya, No.9 (1965), 124–129. (Russian)

GROSSGEIM, V.A., G.B. ARISTOVA, T.D. BAZANOVA  et al.  Problems in Studying and Prospecting for Stratigraphic and Lithologic Oil and Gas Pools in the USSR, pp.54–66. — Leningrad, "Nedra," 1967; (Trudy Vses. Nauchno-Issled. Geologo-Razvedoch. Inst., No.259). (Russian)

GUBKIN, I.M.  Science of Petroleum. — Moscow, Otd. Nauchn. Tekhn. Informatsii, 1937 460 pp. (Russian)

GUBKIN, I.M.  Volga-Ural Petroliferous Region (Second Baku). — Moscow, Gostoptekhizdat, 1940, 119 pp. (Russian)

GULYAEVA, L.A.  Geochemistry of Devonian and Carboniferous Deposits of Kuibyshev Region of the Volga . — Moscow, Akad. Nauk SSSR, 1956, 142 pp. (Russian)

GURARI, F.G.  The Paleogeography of the West Siberian Lowlands in Jurassic-Neocomian Times, pp.37–45. Leningrad, Gostoptekhizdat, 1961; (Trudy Sibirsk. Nauch.-Issled. Inst. Geol., Geofiz. i Mineral. Syr'ya, No.14). (Russian)

GUROVA, T.I. and V.P. KAZARINOV.  Lithology and Paleogeography of the West Siberian Lowlands with Reference to the Oil and Gas Prospects. — Moscow, Gostoptekhizdat, 1962, 296 pp. (Russian)

HAMILTON, D.  Wealth from the North Sea. — New Scientist **38** (1968), 112–136.

HEDBERG, H.D.  Continental Margins from the Viewpoint of the Petroleum Geologist. — Bull. Am. Ass. Petrol. Geol. **54**, No.1, 1970.

HODGSON, G.W.  Market Control on the Development of the Athabasca Oil Sands. — Chem. Engng. Prog. Symp. Ser., No.54 (1965), 72–79.

JOHNSON, M.R. and G.R. WHITE.  Louisiana Gulf Coast. — Bull. Am. Ass. Petrol. Geol. **50**, No.6 (1966), 1232–1243.

KALINKO, M.K.  Main Distribution Patterns of Oil and Gas in the Earth's Crust. — Moscow, "Nedra," 1964, 207 pp. (Russian)

KARAGODIN, Yu.N.  Structural Features and Distribution Patterns of Oil and Gas Fields in West Siberia. — In: Geologiya nefti i gaza, pp.24–31. Saratov, Saratovsk. Univ., 1970. (Russian)

KARPINSKII, A.P.  Outline of the Geologic Past of European Russia. — Moscow, Akad. Nauk SSSR, 1947, 208 pp. (Russian)

KAY, M.  North American Geosynclines. (II). — Geol. Soc., 1951.

KAZARINOV, V.P., V.I. BRATOV, T.I. GUROVA, et al. Weathering and Lithogenesis. — Moscow, "Nedra," 1969, 455 pp. (Russian)

KHAIN, V.E. Geotectonic Principles of Oil Exploration. — Baku, Aznefteizdat, 1954, 692 pp. (Russian)

KHAIN, V.E. Main Belts of Oil and Gas Formation of the World. — Vestnik Moskov. Univ., ser. geol., No.1 (1970), 66—71. (Russian)

KHALIMOV, E.M. On Some Distribution Patterns of Oil-Bearing Sand Reservoirs within the Coal-Bearing Series of Southwestern Bashkiria, pp. 67—73. — Ufa, 1960; (Trudy Bashkirsk. Fil. Akad. Nauk SSSR, No.6). (Russian)

KHANIN, A.A. — Oil and Gas Reservoir Rocks and Their Study. — Moscow, "Nedra," 1969, 366 pp. (Russian)

KHRAMOV, A.N. Paleomagnetic Correlation of Sedimentary Strata. — Leningrad, Gostoptekhizdat, 1958, 219 pp.; (Trudy Vses. Nauch.-Issled. Geologo-Razvedoch. Inst., No.116). (Russian)

KIDWELL, A.L. et al. Oil Migration in Young Deposits of the Padernales Area in Venezuela (Russian translation, 1961).

KING, C.A. Beaches and Coasts, p.434. — London, 1959.

KING, R.E. Prospects of Considerable Increase in World Oil and Gas Reserves by Prospecting for Stratigraphic Traps. — In: Stratigraficheskie lovushki, pp. 3—17. Moscow, Natsional'nyi Komitet SSSR po Nefti, 1971. (Russian)

KLENOVA, M.V. Marine Geology. — Moscow, Uchpedgiz, 1948, 495 pp. (Russian)

KNEBEL, G.M. and G. RODRIGUES-ERASO. Habitat of Some Oil. — Bull. Am. Ass. Petrol. Geol. **40**, No.4 (1956), 547—561.

KRASHENINNIKOV, G.F. Conditions of Accumulation of the Coal-Bearing Formations of the USSR. — Moscow, Moskov. Univ., 1957, 294 pp. (Russian)

KRASHENINNIKOV, G.F. Paleodeltas in the USSR and Certain Aspects of Their Study. — In: Del'tovye i melkovodno-morskie otlozheniya, pp. 7—13. Moscow, Akad. Nauk SSSR, 1963. (Russian)

KRASHENINNIKOV, G.F. Achievements, Present State and Problems of Facies and Paleogeographic Analysis. — In: Sostoyanie i zadachi sovetskoi litologii, pp. 43—57. Moscow, "Nauka," 1970. (Russian)

KROTOVA, V.A. Hydrogeologic Factors in the Formation of Oil Deposits. — Leningrad, Gostoptekhizdat, 1962, 329 pp.; (Trudy Vses. Nauch.-Issled. Geologo-Razvedoch. Inst., No.191). (Russian)

KUENEN, Ph.H. Sand: Its Origin, Transportation, Abrasion and Accumulation. — Geol. South Afric., No.6 (1959), 193—204.

KUZNETSOV, S.I., M.V. IVANOV, and N.N. LYALIKOVA. Introduction to Geologic Microbiology. — Moscow, Akad. Nauk SSSR, 1962, 239 pp. (Russian)

KUZNETSOV, V.G. Principal Aspects of the Geology of Reefs and Their Oil and Gas Prospects. — Moscow, Vses. Nauch.-Issled. Inst. Organizatsii Upr. i Ekonom. Neftegaz. Promyshlennosti, 1971, 60 pp. (Russian)

LEONT'EV, O.K. A Brief Course in Marine Geology. — Moscow, Moskov. Inst., 1963, 464 pp. (Russian)

LEVORSEN, A.I. Geology of Petroleum. Edited by Berry, F.A. 2nd Edition. — W.H. Freeman, 1967.

LOGVINENKO, N.V. Lithology and Paleogeography of the Carboniferous Productive Series of the Donbass. — Kharkov, Kharkovsk. Univ., 1959, 436 pp. (Russian)

LOWMAN, S.W. Sedimentary Facies in the Gulf Coast. — Bull. Am. Ass. Petrol. Geol. **33**, No.12 (1949), 1939—1997.

MAKSIMOV, S.P. Distribution Patterns and Conditions of Formation of Oil and Gas Pools. — Moscow, "Nedra," 1964, 485 pp. (Russian)

MARKOV, K.K. Paleogeography. — Moscow, Moskov. Univ., 1960, 268 pp. (Russian)

MARKOVSKII, N.I. The Lower Visean Paleogeography of the Middle Volga and Transvolga Regions. — Dokl. Akad. Nauk SSSR **104**, No.4 (1955), 601—604. (Russian)

MARKOVSKII, N.I. Coal Prospects of the Lower Carboniferous in Middle Volga and Transvolga Regions, pp. 366—378. — Moscow, Akad. Nauk SSSR, 1956; (Trudy Labor. Uglya Akad. Nauk SSSR, No.6). (Russian)

MARKOVSKII, N.I. Coal and Oil in the Volga-Ural Region. — Priroda, No.5 (1957), 88—91. (Russian)

MARKOVSKII, N.I. On the Relationship between Oil Source Rocks and Coal-Bearing Formations of the Volga-Ural Region. — Geologiya Nefti i Gaza, No.3 (1959), 22—29. (Russian)

MARKOVSKII, N.I. Occurrence of Coal and Prediction of the Occurrence of Oil and Gas. — Priroda, No.4 (1959), 89—93. (Russian)

MARKOVSKII, N.I. Paleogeographic Situation of the Kizel Coal Basin and Its Genetic Characteristics in the Light of Recent Data. — Izv. Akad. Nauk SSSR, ser. geol., No.2 (1960), 65—76. (Russian)

MARKOVSKII, N.I., G.A. BRAZHNIKOV, and T.L. VESNINA. On Surveying Lithologic Oil and Gas Pools in the Volga Region Near Volgograd. — Geologiya Nefti i Gaza, No.4 (1962), 25—29. (Russian)

MARKOVSKII, N.I.  Paleogeography and the Search for Rich Oil Pools. — Priroda, No.8 (1963), 88—90.  (Russian)

MARKOVSKII, N.I.  The Role of Paleorivers in the Formation of Oil- and Gas-Bearing Series. — Izv. Akad. Nauk SSSR, ser. geol., No.2 (1965), 112—125.  (Russian)

MARKOVSKII, N.I.  Paleogeographic Factors Determining the Disposition of Large Oil Pools. — Moscow, "Nedra," 1965, 399 pp.  (Russian)

MARKOVSKII, N.I.  Zoned Distribution of Oil and Gas Fields. — Moscow, Vses. Nauch.-Issled. Inst. Organizatsii Upr. i Ekonom. Neftegaz. Promyshlennosti, 1967, 104 pp.  (Russian)

MARKOVSKII, N.I.  Paleogeographic Assessment of Belts and Major Zones of Oil and Gas Accumulation. — Izv. Akad. Nauk SSSR, ser.geol., No.10 (1968), 130—138.  (Russian)

MARKOVSKII, N.I.  Ancient Coastal Zones and the Distribution of Oil and Gas Deposits. — Byull. Moskov. Obshch. Ispyt. Prir., otd. geol., No.4, 1969, 143 pp.  (Russian)

MARKOVSKII, N.I.  Paleodeltas — Zones of Maximum Accumulation of Coal, Oil and Gas. — In: Genezis i klassifikatsiya osadochnykh porod, pp.154—161.  Moscow, "Nauka," 1968; (Dokl. Sovet. Geologov na 23-ei Sessii Mezhduvedomstv. Geofiz. Komiteta).  (Russian)

MARKOVSKII, N.I.  The Role of Paleogeographic Studies in Ascertaining the Distribution Patterns of Oil and Gas Fields, pp.128—130. — Moscow, Moskov. Univ., 1969; (Sb. 4-oi Nauch. Konf. Geol. Fak. Moskov. Gos. Univ.).  (Russian)

MARKOVSKII, N.I.  Paleogeographic Zonal Distribution of Oil and Gas in West Siberia. — In: Prirodnye resursy Zapadnoi Sibiri, pp.48—64.  Moscow, Moskov. Gos. Univ., 1971.  (Russian)

MARKOVSKII, N.I.  Much Gas in the Yamal-Nenets Territory. — Priroda, No.8 (1971), 19—24.  (Russian)

MARTINI, I.P.  Regional Analysis of the Sedimentology of the Medina Formation. — Bull. Am. Ass. Petrol. Geol. **55**, No.8 (1971), 1249—1261.

MASLOV, K.S.  On Prospects of Exploration for Lithologic Oil Pools in Lower Maikop Deposits of the Northwestern Caucasus. — Geologiya Nefti i Gaza, No.4 (1962), 29—35.  (Russian)

MASLOV, K.S.  Scientific Principles of Exploration for Lithologic and Stratigraphic Oil and Gas Pools in Terrigenous Series, p.220. — Moscow, "Nedra," 1968.  (Russian)

MATHEWS, W.H. and F.P. SHEPARD.  Sedimentation of the Fraser River Delta, British Columbia. — Bull. Am. Ass. Petrol. Geol. **46**, No.8 (1962), 1416—1437.

Methods of Preparing Lithofacies and Paleogeographic Maps, p.176. — Novosibirsk, Sibirsk. Otd. Akad. Nauk SSSR, 1963; (Trudy 5-go Vses. litolog. soveshch.).  (Russian)

Methods of Paleogeographic Studies, p.264. — Moscow, "Nedra," 1964; (Sb. Mater. Vyezdn. Sessii Eksp. Geol. Soveta Geolkoma).  (Russian)

MIKHAILOV, V.N.  Some Features of the Formation of Bars and River Mouths, Considerably Influenced by River Action, pp.67—75. — Moscow, "Nauka," 1967; (Trudy Okeanografich. Inst., No.89).  (Russian)

MOLCHANOV, V.I. and A.A. GONTSOV.  Water-Soluble Organic Compounds as Source Material for Oil Formation. — Dokl. Akad. Nauk SSSR **191**, No.3 (1970), 681—683.  (Russian)

MURRAY, G.E.  Geologic Framework of the Gulf Coast Province of the United States. — In: Rec. Sedim. N-W Gulf of Mexico; Bull. Am. Ass. Petrol. Geol., pp.5—33, 1960.

MURRAY, G.E.  Petroliferous Carbonate Bank in Upper Devonian Strata.  (Russian translation, 1968)

NAIDIN, D.P.  Determination of Climatic Conditions during the Geologic Past Using the Method of Isotope Paleothermometry. — Sovetskaya Geologiya, No.7 (1958), 15—34.  (Russian)

NAIRN, A.E.M.  Problems in Paleoclimatology. — New York, Wiley (Interscience), 1964.

NALIVKIN, D.V.  Theory of Facies, Vols.1 and 2, pp.534 and 393. — Moscow, Akad. Nauk SSSR, 1956.  (Russian)

NALIVKIN, V.D., A.B. RONOV, and V.E. KHAIN.  Basic Principles of Preparation of Lithofacies and Paleogeographic Maps of the East European Platform and Its Geosynclinal Margin, pp.25—41. — Novosibirsk, Sibirsk. Otd. Akad. Nauk SSSR, 1963; (Trudy 5-go Vses. Litol. Soveshch., Vol.1).  (Russian)

NALIVKIN, V.D. and K.A. CHERNIKOV (Editors).  Formation Conditions and Distribution Patterns of Oil and Gas Fields. — Moscow, "Nedra," 1967, 296 pp.  (Russian)

NALIVKIN, V.D., V.N. VERESHCHAGIN, and G.S. GANESHIN.  Annotations and Methodology in Preparing the Atlas of Lithopaleogeographic Maps of the USSR. — Moscow, Gosgeoltekhizdat, 1962, 46 pp.  (Russian)

NERUCHEV, S.G., M.F. DVALI, V.A. KROTOVA et al.  Exploration Criteria for Prediction of Oil and Gas Occurrences. — Leningrad, "Nedra," 1969; (Trudy Vses. Nauch.-Issled. Geologo-Razvedoch. Inst., No.269).  (Russian)

NESTEROV, I.I.  Criteria for the Prediction of Oil and Gas Occurrences, p.335. — Moscow, "Nedra," 1969; (Trudy Zap.-Sibirsk. Nauch.-Issled. Geologo-Razvedoch. Neft. Inst., No.15).  (Russian)

POTTER, D.E.  Sand Bodies and Sedimentary Environments, a Review. — Bull. Am. Ass. Petrol. Geol. **51**, No.3 (1967), 337—365.

RAINWATER, E.H.  Look for Ancient Deltas in Your Search for Oil. − Oil Gas. J. **62**, No.1 (1964), 42−57.

RODIONOVA, K.F.  The Significance of Facies Conditions in the Formation of Oil, pp.18−30. − Leningrad, Gostoptekhizdat, 1959; (Trudy Vses. Neft. Nauch.-Issled. Geologo-Razvedoch. Inst., No.17). (Russian)

RUKHIN, L.B.  Principles of General Paleogeography. − Moscow, Gostoptekhizdat, 1959, 557 pp. (Russian)

RUKHIN, L.B.  Principles of Lithology. − Moscow, Gostoptekhizdat, 1961, 779 pp. (Russian)

RUSSELL, R.J.  Physiography of the Lower Mississippi Delta. − Geol. Bull., No.8 (1936), 3−99.

RUSSELL, W.L.  Principles of Petroleum Geology. − McGraw-Hill, 1951.

SAMOILOV, I.V.  River Mouths. − Moscow, Geografgiz, 1952, 526 pp. (Russian)

SARKISYAN, S.G., M.V. KORZH, G.N. KOMARDINKINA et al.  West Siberia in the Jurassic Period. − Moscow, "Nauka," 1967, 160 pp. (Russian)

SARKISYAN, S.G. and T.N. PROTSVETAEVA.  Paleogeography of the West Siberian Lowlands during the Early Cretaceous. − Moscow, "Nauka," 1968, 80 pp. (Russian)

SARKISYAN, S.G. and G.N. KOMARDINKINA.  Lithofacies Features of the Cenomanian Gas-Bearing Deposits in the North of the West Siberian Lowlands. − Moscow, "Nedra," 1971, 115 pp. (Russian)

SCOTT, A.J. and W.L. FICHER.  Delta Systems and Deltaic Deposition. − Texas Univ. Bur. Econ. Geol., No.10 (1969), 10−29.

SHANNON, J.P. and A.R. DAHL.  Deltaic Stratigraphic Traps in West Tuscola Field, Taylor County, Texas. − Bull. Am. Ass. Petrol. Geol. **55**, No.8 (1971), 1194−1205.

SHANTSER, E.V.  Alluvium of Plain Rivers of the Temperate Zone and Its Significance in the Study of the Regularities Governing the Structure and Formation of Alluvial Suites. − Moscow, Akad. Nauk SSSR, 1951, 275 pp.; (Trudy Inst. Geol. Nefti Akad. Nauk SSSR, No.135). (Russian)

SHANTSER, E.V.  Outline of the Theory of Genetic Types of Continental Sedimentary Formations. − Moscow, "Nauka," 1966, 239 pp. (Russian)

SHEPARD, F.P.  Earth Beneath the Sea. − New York, Atheneum, 1964.

SHEPARD, F.P.  Submarine Geology. 2nd Edition. − New York, Harper and Row Publishers, 1963.

SHEPARD, F.P. and R.A. LANKFORD.  Sedimentary Facies from Shallow Borings in the Lower Mississippi Delta. − Bull. Am. Ass. Petrol. Geol. **43**, No.9 (1959), 989−998.

SILIN-BEKCHURIN, A.I.  Dynamics of Groundwaters. − Moscow, Moskov. Univ., 1958, 258 pp. (Russian)

SMIRNOV, G.A.  Material on Paleogeography of the Urals, Visean Stage. − Sverdlovsk, 1957, 118 pp.; (Trudy Gorn. Geol. Inst. Uralsk. Fil. Akad. Nauk SSSR, No.29). (Russian)

SMITH, J.  Basement Reservoir of La Paz-Mara Oil Fields, Western Venezuela. − Bull. Am. Ass. Petrol. Geol. **40**, No.2 (1956), 310−318.

SOROKIN, Yu.I.  A Voyage to the Coral Islands. − Priroda, No.8 (1971), 61−68. (Russian)

STEPANOV, P.I.  Theory of Belts and Major Zones of Coal Accumulation. − Yubil. Sb. posvyashch. 30-letiyu Oktyabr. rev., Part 2, pp.172−193. Moscow, Akad. Nauk SSSR, 1947. (Russian)

STRAKHOV, N.I., E.S. ZALMANZON, and M.A. GLAGOLEVA.  Outline of the Geochemistry of Upper Paleozoic Deposits of the Humid Type. − Moscow, Akad. Nauk SSSR, 1959, 223 pp.; (Trudy Geol. Inst. Akad. Nauk SSSR, No.23). (Russian)

STRAKHOV, N.M.  Types of Climatic Zones in the Post-Proterozoic History of the Earth and Their Geological Importance. − Izv. Akad. Nauk SSSR, ser. geol., No.3 (1960), 3−25. (Russian)

STRAKHOV, N.M.  Fundamentals of the Theory of Lithogenesis, Vols.1 and 2, pp.212 and 574. − Moscow, Akad. Nauk SSSR, 1962. (Russian)

STRAKHOV, N.M.  On the True Balance of Sedimentation in the Present-Day Ocean. − Litologiya i Poleznye Iskopaemye, No.2 (1969), 131−132. (Russian)

STRAKHOV, N.M.  Quantitative Methods of Studying Sedimentation in the Geologic Past. − Litologiya i Poleznye Iskopaemye, No.3 (1970), 3−14. (Russian)

STRAKHOV, N.M.  Evolution of Lithogenetic Concepts in Russia and in the USSR. − Moscow, "Nauka," 1971, 609 pp. (Russian)

SWANN, D.H.  Late Mississippian Rhythmic Sediments of the Mississippi Valley. − Bull. Am. Ass. Petrol. Geol. **48**, No.5 (1964), 637−658.

TEICHMÜLLER, M.  Anwendung kohlenpetrographischer Methoden bei der Erdöl- und Erdgasprospektion. − Erdöl Kohle Erdgas Petrochem. **24**, No.2 (1971), 69−76.

TEODOROVICH, G.I. and M.V. BOGDASAROVA.  The Transition from Coal-Bearing Deposits to Oil- and Gas-Producing Deposits. − Geologiya Nefti i Gaza, No.8 (1967), 42−47. (Russian)

THOMAS, T.M.  The North Sea and Its Environs. − Future Reservoir of Fuel. − Geol. Rep. **56**, No.8 (1966), 1789−1796.

TIMOFEEV, P.P.  Alluvial Deposits and Associated Erosion in the Middle Carboniferous Coal-Bearing Suites on the Southwestern Margin of the Donbass. — In: Allyuvial'nye otlozheniya v uglenosnoi tolshche srednego karbona Donbassa, pp.90—116.  Moscow, Akad. Nauk SSSR, 1964; (Trudy Inst. Geol. Nefti Akad. Nauk SSSR, No.151). (Russian)

TIMOFEEV, P.P.  The Jurassic Coal-Bearing Formation of South Siberia and the Conditions under Which it Was Formed. — Moscow, "Nauka," 1970, 208 pp. (Russian)

URUPOV, A.K.  On Mapping Regions with Increased Thickness of Lower Carboniferous Terrigenous Formations by the Seismic Reflection Method. — Geologiya Nefti i Gaza, No.2 (1961), 29—31. (Russian)

USPENSKAYA, N.Yu.  Certain Features of Oil and Gas Accumulations on Platforms. — Moscow, Gostoptekhizdat, 1952, 156. (Russian)

USPENSKII, V.A.  Introduction to Petroleum Geochemistry. — Moscow, "Nedra," 1970, 309 pp. (Russian)

Van KREVELEN, D.W. and J. SCHUVER.  Coal Science, p.223. — Amsterdam-London-New York, 1957.

VASIL'EV, P.V.  Paleogeographic Conditions of Formation of Lower Carboniferous Coal-Bearing Deposits on the Western Slope of the Urals. — Moscow, Ugletekhizdat, 1950, 208 pp. (Russian)

VASIL'EV, V.G. and A.A. KHANIN.  Distribution of Oil and Gas Deposits in the Section Across the Sedimentary Cover of the USSR. — Geologiya Nefti i Gaza, No.11 (1963), 1—5. (Russian)

VASSOEVICH, N.B.  Formation of Oil in Terrigenous Deposits, p.220. — Leningrad, Gostoptekhizdat, 1958; (Trudy Vses. Nauch.-Issled. Geologo-Razvedoch. Inst., No.128). (Russian)

VASSOEVICH, N.B.  Theory of Sedimentary-Migrational Origin of Petroleum. — Izv. Akad. Nauk SSSR, ser. geol., No.11 (1967), 135—156. (Russian)

VASSOEVICH, N.B., I.V. VYSOTSKII, A.N. GUSEVA, and V.B. OLENIN (Authors).  Hydrocarbons in the Earth's Sedimentary Mantle. — Vestnik Moskov. Univ., ser.geol., No.5 (1967), 36—48. (Russian)

VASSOEVICH, N.B., Yu.M. KORCHAGINA, N.V. LOPATIN, and V.V. CHERNYSHEV.  The Main Phase in Oil Formation. — Vestnik Moskov. Univ., ser.geol., No.6 (1969), 3—27. (Russian)

VEBER, V.V.  Facies Deposits Conducive to Oil Formation. — Moscow, "Nedra," 1966, 273 pp. (Russian)

VEBER, V.V., T.L. GINZBURG-KARAGICHEVA, and E.A. GLEBOVSKAYA.  Accumulation and Transformation of Organic Matter in Recent Marine Sediments. — Moscow, Gostoptekhizdat, 1956, 343 pp. (Russian)

VERNADSKII, V.I.  The Biosphere. — Moscow, "Mysl'," 1967, 376 pp. (Russian)

VER WIEBE, W.A.  How to Find Oil? — Moscow, Gostoptekhizdat, 1959, 275 pp. (Russian)

VISHER, G., B. SAITTA, and R. PHARES.  Pennsylvanian Delta Patterns and Petroleum Occurrences in Eastern Oklahoma. — Bull. Am. Ass. Petrol. Geol. 55, No.8 (1971), 1206—1229.

VYSOTSKII, I.V. and V.B. OLENIN.  Deep Zoning and Distribution of Hydrocarbon Accumulations. — Vestnik Moskov. Univ., ser. geol., No.6 (1964), 20—27. (Russian)

WEAVER, P.  Gulf of Mexico. — Spec. Pap. Geol. Soc. Am., No.62 (1955), 269—278.

WEBER, K.J.  Sedimentology of Oil Fields in the Niger Delta. — Geol.-mijnbouwk. Biblphie Ned.-Indie 50, No.3 (1971), 559—576.

WEIMER, R.  Deltas and Petroleum. — Bull. Am. Ass. Petrol. Geol. 55, No.8 (1971), 1135—1136.

YOUNG, T.A.  Prospecting for Stratigraphic Traps in the Oficina Area (Venezuela). — In: Stratigraficheskie lovushki, pp.63—81.  Moscow, Natsional'nyi Komitet po Nefti, 1971. (Russian)

ZENKEVICH, L.A.  Seas of the USSR, Their Fauna and Flora. — Moscow, Uchpedgiz, 1955, 424 pp. (Russian)

ZENKOVICH, V.P.  Theory of Seashore Development. — Moscow, Akad. Nauk SSSR, 1962, 720 pp. (Russian)

ZHEMCHUZHNIKOV, Yu.A.  On the Possible Presence and Conditions of Burial of Alluvial Deposits in Fossil Strata. — In: Allyuvial'nye otlozheniya v uglenosnoi tolshche srednego karbona Donbassa, pp.9—29. Moscow, 1954; (Trudy Geol. Inst. Akad. Nauk SSSR, No.151). (Russian)

ZHEMCHUZHNIKOV, Yu.A.  Coal-Bearing Strata as Formations. — Izv. Akad. Nauk SSSR, ser. geol., No.5 (1955), 46—58. (Russian)

ZHIZHCHENKO, B.P.  Methods of Paleogeographic Investigation. — Moscow, Gostoptekhizdat, 1959, 371 pp. (Russian)

ZHIZHCHENKO, B.P.  Sedimentation Conditions and Methods of Paleogeographic Reconstructions. — Sovetskaya Geologiya, No.9 (1965), 47—62. (Russian)

# INDEX

Science

# Date Due

| | | | |
|---|---|---|---|
| | | | |
| | | | |
| | | | |
| | | | |
| | | | |
| | | | |
| | | | |
| | | | |
| | | | |
| | | | |
| | | | |
| | | | |
| | | | |
| | | | |
| | | | |
| | | | |
| | | | |
| | | | UML 735 |